Online Social Networking

Other Books in the Current Controversies Series

Current CONTROVERSIES

Online Social Networking

Sylvia Engdahl, Book Editor

GREENHAVEN PRESS

An imprint of Thomson Gale, a part of The Thomson Corporation

THOMSON

GALE

Detroit • New York • San Francisco • New Haven, Conn. • Waterville, Maine • London

THOMSON

GALE

Printed in the United States of America
10 9 8 7 6 5 4 3 2 1

Contents

No: Online Social Networking Is Beneficial to Teens

Chapter 2: Should Congress Require Schools and Libraries to Block Networking Sites?

Chapter 3: Who Is Using Online Social Networking?

Chapter 4: How Is Online Social Networking Changing Society?

Foreword

By definition, controversies are "discussions of questions in which opposing opinions clash" (Webster's Twentieth Century Dictionary Unabridged). Few would deny that controversies are a pervasive part of the human condition and exist on virtually every level of human enterprise. Controversies transpire between individuals and among groups, within nations and between nations. Controversies supply the grist necessary for progress by providing challenges and challengers to the status quo. They also create atmospheres where strife and warfare can flourish. A world without controversies would be a peaceful world; but it also would be, by and large, static and prosaic.

The Series' Purpose

The purpose of the Current Controversies series is to explore many of the social, political, and economic controversies dominating the national and international scenes today. Titles selected for inclusion in the series are highly focused and specific. For example, from the larger category of criminal justice, Current Controversies deals with specific topics such as police brutality, gun control, white collar crime, and others. The debates in Current Controversies also are presented in a useful, timeless fashion. Articles and book excerpts included in each title are selected if they contribute valuable, long-range ideas to the overall debate. And wherever possible, current information is enhanced with historical documents and other relevant materials. Thus, while individual titles are current in focus, every effort is made to ensure that they will not become quickly outdated. Books in the Current Controversies series will remain important resources for librarians, teachers, and students for many years.

In addition to keeping the titles focused and specific, great care is taken in the editorial format of each book in the series. Book introductions and chapter prefaces are offered to provide background material for readers. Chapters are organized around several key questions that are answered with diverse opinions representing all points on the political spectrum. Materials in each chapter include opinions in which authors clearly disagree as well as alternative opinions in which authors may agree on a broader issue but disagree on the possible solutions. In this way, the content of each volume in Current Controversies mirrors the mosaic of opinions encountered in society. Readers will quickly realize that there are many viable answers to these complex issues. By questioning each author's conclusions, students and casual readers can begin to develop the critical thinking skills so important to evaluating opinionated material.

Current Controversies is also ideal for controlled research. Each anthology in the series is composed of primary sources taken from a wide gamut of informational categories including periodicals, newspapers, books, U.S. and foreign government documents, and the publications of private and public organizations. Readers will find factual support for reports, debates, and research papers covering all areas of important issues. In addition, an annotated table of contents, an index, a book and periodical bibliography, and a list of organizations to contact are included in each book to expedite further research.

Perhaps more than ever before in history, people are confronted with diverse and contradictory information. During the Persian Gulf War, for example, the public was not only treated to minute-to-minute coverage of the war, it was also inundated with critiques of the coverage and countless analyses of the factors motivating U.S. involvement. Being able to sort through the plethora of opinions accompanying today's major issues, and to draw one's own conclusions, can be a complicated and frustrating struggle. It is the editors' hope that Current Controversies will help readers with this struggle.

Introduction

Over the past few years a great deal of media attention has been given to online social networking, especially networking Web sites heavily used by teens, such as MySpace and Facebook. Many people believe that the growing use of these sites is causing changes in the nature of social relationships that will, perhaps, have a profound effect on society. The question of whether this new form of relationship is good or bad has become increasingly controversial.

Most people assume that online networking, apart from business contexts, did not exist until quite recently, and that its impact on how people relate to each other is only now being discovered. Actually, such networking is not new; it began in the 1980s, before the existence of the Internet, and many of the people involved in it at that time were interested in analyzing its significance. An article published in the November 1985 issue of *Netweaver*, the online newsletter of the Electronic Networking Association, listed fourteen benefits of what was then called "computer conferencing," many of which are characteristics that are still being noticed today. Among other facts, it mentioned that each individual can communicate with more people via CC (computer conferencing) than in any other way; that in CC, people of all ages and backgrounds meet on equal footing; that geography is no longer relevant to communication; that people can communicate at their own convenience (maybe even in the middle of the night); that CC brings people with common interests together; and—most notably—that there is less inhibition in CC than in most face-to-face conversations, simply because its social conventions permit people who are not intimately acquainted to discuss their inner feelings in ways not customary at social gatherings. The article also noted that when people who have met online meet in person, they start off at a higher level of involvement than they would if they were strangers.

However, although all these things were recognized in the 1980s, they affected only a small group of people—the minority who had home computers and, in most cases, a good deal of money to spend on telecommunications. In those days, it was not possible to hook up to a national network for a low monthly fee; time online was charged by the minute at a high rate, and modems were incredibly slow by today's standards. Furthermore, there were a number of different networks, none of which were connected to each other; people could communicate only with others who had accounts on the same one. The main barrier to the kind of future enthusiasts then envisioned—which has now arrived—was cost; it was felt that unless some sort of tax-supported network could be established, only affluent people would be able to participate. The World Wide Web, supported to a large extent by advertising, was not foreseen.

There was, to be sure, one form of online communication popular among teens who owned computers in the 1980s: local electronic bulletin board systems, or BBSs. These systems were free; they were operated by hobbyists—who were in many cases teens themselves—and functioned not merely for the posting of "bulletins" but for discussion among friends. Many of these friends had not met in person, but in other cases, they saw each other in school yet also communicated online, just as teens do today. BBSs, which ran on computers that were primitive compared to modern models, were limited to a few hundred users each, all of whom lived close enough to phone in without incurring long-distance charges. Large cities had dozens of separate ones. Nevertheless, people who used them often experienced the same sense of community that was characteristic of participants in national conferencing systems, and which is still among the most prominent features of online social networking.

In 1985 on The Source, one of the earliest national conferencing systems, an adult member wrote: "I think what really hooked me was the effect I saw CC having on *me*. Within the first few days after I began participating I began to feel connectedness with those with whom I was corresponding that I seldom feel, even with people I have known and worked with for years. I also found telecommunicating to be profoundly disinhibiting. My thoughts seemed to flow in an easy way, and I felt that that which I was saying was somehow more heartfelt than usual." Over and over during the ensuing years, the same thoughts have been expressed. There is something inherent in online social networking that causes people to reveal more of themselves than they do in any other situation.

Back in the 1980s, this did not create trouble except when someone made frank but untactful remarks that caused dissension. But with the Internet and low-cost access, which opened the world of online networking to virtually everyone—at public libraries if not at home—serious problems began to arise. What people say online can now be read not only by friends and potential friends, but by teachers, college admission officials, and employers. Worse, it can be read by sexual predators and other criminals, who prowl the Web looking for victims. And so a dark side of online social networking has developed. This has caused some people to feel that the risks of participation, especially for young teens, outweigh the benefits. Others believe that the risks are no greater than those of other contact situations that children have always been taught to watch out for.

Additional aspects of social networking are disturbing to some. Those who value personal privacy are dismayed to see that today it is being willingly abandoned by new generations, and in some cases by older ones. Also, some observers fear that young people who do most of their communication online will never develop the skills needed for face-to-face communication, and that a reduction in face-to-face communication would be bad for society. Not everyone thinks that changes in ways of communicating are bad, however. Historically, there have been many such changes. When the printing press was invented, some people deplored the fact that it was likely to lead to a decline in the ability to memorize long stories and pass them on orally, which in fact it did; but few today consider that a significant loss.

If online social networking was not a new idea at the time Internet access became common in the late 1990s, why did its use accelerate so much in 2005 and 2006, the years in which it first became a widely discussed public issue? There were plenty of forums for communication on the Internet previously—for example, Usenet groups—and many were popular. Yet they did not attract the vast numbers of users that MySpace does, or play a major role in the lives of large numbers of average young people.

There are two reasons for the dramatic rise in social networking. In the first place, sites such as MySpace are viewed as *places*, not mere message boards—users think of these locations in cyberspace as if they were physical gathering places. Before they existed, the potential of the Internet to bring together large numbers of people was not realized because although there were many Web sites where people could talk, only a relatively small number visited each of them. Most online communities were not much larger in terms of ongoing participants than BBS groups had been, and they were less cohesive because of their openness to a virtually infinite number of transient users whose existence was revealed only through scattered comments. Unlike such sites, the major new ones offer individuals a permanent presence. The main factor in the new sites' drawing power, however, is their multimedia capability. For the first time in social networking, participants are able to post pictures as well as text. Technologically skilled people previously posted pictures at their own Web sites, but comparatively few viewers found those sites. MySpace and similar sites have made it easy for unskilled users to create multimedia pages, as well as to discover the pages of friends old and new. The majority of social networkers still communicate with only a small subset of the millions of people who have pages, but those millions are nevertheless perceived as belonging to a neighborhood—a neighborhood as real, to many, as the city in which they live their offline lives.

Because the freer and less inhibited self-expression inherent in online communication has no offline substitute, it is not surprising that so many young people favor it over more traditional forms of socialization. "It's a very sad testimonial these days that a kid has to post something on a site where potentially 700 million people can see it in order to attract the attention of a kid two seats down," Internet safety expert Perry Aftab told the *Washington Post*. "They do less face-to-face talking, less phone talking, less playing outside than any other generation, and because of that, the Internet is real to them, but the risks aren't." It is indeed unfortunate when teens ignore the risks, and no one doubts that better education is needed to make them observe safety precautions. However, it is debatable whether the trend toward replacing face-to-face and phone talk with online communication is "sad."

In any case, since the disinhibiting—and in some cases, addictive—nature of online communication is a universal and long-standing phenomenon that has been recognized by participants for more than twenty years, it is unlikely that online social networking will turn out to be a mere fad. For better or for worse, it is becoming a way of life for a growing proportion of the world's population.

CHAPTER 1

Should Parents Limit Teens' Use of Social Networking Web Sites?

Chapter Preface

To many teens, Web sites such as MySpace and Facebook are an integral part of daily life, one they cannot imagine being without. It seems so natural to them to use these sites for communication with their friends that they find it hard to understand why social networking is controversial. To parents, however, what can be found on MySpace is apt to be horrifying. They object not merely to the physical risks involved in displaying private information that might be seen by criminals—which, though all too real, affect relatively few of the site's users—but to the much larger risk that adolescents will be influenced by inappropriate material posted by other young people.

All kinds of people have MySpace profiles, including pornographers and teens who lack the maturity to distinguish between innocent posing and porn. There are also teens who know perfectly well what constitutes acceptable behavior; some pretend to have done things they would never actually do in person, while others provide true accounts of their misconduct. Understandably, since most parents would not allow their children to engage in sex, drinking, and so forth in their homes, they are reluctant to have them exposed to such activities online. The mere fact that the display of these things is accessible seems threatening, as does the prevalence of crude language in many profiles. Therefore, some parents feel that teens should not be permitted to use social networking sites— even those teens who generally display sound judgment.

Others, however, believe that teens need to learn how to interact in society and that social networking is merely the modern form of an extended neighborhood. It does offer contact with undesirable role models, but it also provides a wide variety of valuable contacts to whom teens might otherwise not be exposed. Only through experience, they maintain, will young people gain the ability to judge what they are likely to meet in the connected world of tomorrow.

Experts are divided. All agree that the use of social networking sites should be overseen by parents and that teens should be taught never to post revealing personal information. Many law enforcement officials consider the danger of being solicited by sexual predators large enough to warrant keeping off MySpace entirely. However, some educators and psychologists point out that social networking can be beneficial to teens. "MySpace and other sites open up a world where they can test out who they are and who they want to be," says Debbie Beach, a therapist who specializes in treating adolescents. It also gives them an opportunity to be creative in the design of their pages and in what they write; not all teens confine their online comments to mere trivia.

"Just how dangerous is the unsupervised use of the Internet by adolescents? Nobody knows," Caitlin Flanagan wrote in the July 2007 issue of the *Atlantic Monthly*. "When something new comes along, it takes a while for parents to sort out what's safe and what isn't, and even longer for their conclusions to become commonly held assumptions about good parenting. In the future it may be unheard-of for a teenage girl from a loving family to disappear into her room every night for two hours of unsupervised e-chatting and instant messaging and MySpacing. Then again, it may be even more common than it is today. All we know for sure is that our children are living in the midst of a technological revolution, and that they're drawn to it like moths to a flame."

MySpace Encourages Teen Involvement with Porn

Rebecca Hagelin

Rebecca Hagelin is a vice president of the Heritage Foundation, a conservative education and research organization. She is a national columnist and speaks frequently on radio and television.

MySpace.com has quickly become the malt shop for today's teens—but, unlike Al's place in "Happy Days," this virtual "hang out" is also frequented by unsavory characters who are after our kids. What most teens see as just a fun place to connect with friends has become a sexual predator's and pornographer's playground.

As the *New York Times* reported on May 27, the site's content has become so skanky that even some corporate advertisers, including Weight Watchers and T-Mobile, are either pulling their ads from MySpace or threatening to do so.

The pages of sex-film stars like Jenna Jameson and Tera Patrick are quickly becoming popular with America's teens as evidenced by the thousands of "friends"—other MySpace users who ask for full access to a user's page. Those pages, in turn, link to the stars' own homepages, which are often replete with explicit pornographic images.

MySpace Profiles Link to Porn Sites

Some of the stars' "friends" are as young as 14. Others are undoubtedly younger—one need only state their age, no ID or proof required. Most of the linked porn sites post a simple warning that says something like, "Click here to enter if you're over 18," and that's it.

And if it wasn't bad enough having porn stars market themselves directly to your children, try this: Some of the

Rebecca Hagelin, "Porn, Pedophiles, Our Kids and Myspace.com," *ChristianWorld viewNetwork.com*, May 29, 2006. Copyright © 2006 *WorldviewWeekend.com*. Reproduced by permission.

morally depraved individuals who operate behind the camera post profiles on MySpace as well—and they actually encourage girls to send them pictures and contact them about getting involved in the porn business.

Take Eon McKai, a Los Angeles man who works for VCA, a porn company owned by Hustler. In his MySpace profile, he admits, "I make smut" and puts this under the heading "Who I'd like to meet" (typos in the original):

> Cool people not kids but young adults or poeple who fell like young adults that are open minded and like porn and want to be apart of my movement to make it better . . . If you want to be a porn star . . . I'm the guy that can make that happen for real . . . but I'll look you in the eye first to make sure you can hang . . . any Emo, Punk or Goth kids . . . your all my friends . . . be stylish and cute as all hell.

As with most MySpace profile pages, McKai has many "friends" who have posted messages and pictures. One is from a girl named Jess who writes: "hey love! i'm live on my web cam right NOW!! come play with me!!!" Then she lists a porn site that, she notes, is "free for myspace members!"

With the millions of great sites that serve our families well, there's no need for our kids to spend time on a place like MySpace.

Perhaps more disturbing is the number of young women who note in their MySpace profiles that they consider themselves aspiring porn stars. One named Alice in Connecticut writes: "I am shy if you meet me for real but what goes on behind closed doors is a different matter! Just on here looking to meet new My Space people because I was bored and thought it could be a tad interesting . . . just a tad . . . LOL. I've been coming out of my shell lately and doing more modeling and stuff. Even though it is CRAZY, I really AM studying to be a porn star."

Many other profiles on MySpace can't even be quoted here because they are so vile. Even with all the negative media reports about pedophiles hunting kids through MySpace, most parents are still clueless about the content and dangers.

MySpace is just the latest example of the creeping oversexualization of our culture that I describe in my book, *Home Invasion* (written before MySpace became so popular with our youth). The truth is that pornographers aren't content to exist only in their own sick little corner of the Internet: They've got to troll for new victims. And they do it by wading into what amounts to an Internet kiddie pool and snagging innocent kids. That's why your child's e-mail inbox is constantly filled with spam porn. It's why sites such as MySpace have been infiltrated by perverts. It's why you've got to be worried about what's going on within the "safety" of your own home if you have unlimited, unmonitored Internet access.

Filter out MySpace

Parents, vigilance is key. Protecting your kids from Internet smut may be hard, but it's not impossible. Talk to your kids. Keep your computer in a common area. Limit Internet time. Know who your kids are chatting with online. And get an Internet filter–NOW! There are many great, inexpensive filters on the market, but I use the filter from BSafe.com.

When I first started researching the dangers and garbage prevalent on MySpace months ago, I asked my then-13-year-old daughter, "Kristin, do you know about a site called MySpace?" She replied with a sigh, "Yes, Mom. But I can't get to it because our filter blocks it out!" After whispering a prayer of thanks for such a great filter, I asked her, "How many of your friends have pages on MySpace?" She replied, "All of them." What followed, of course, was a serious discussion of why her time was best spent elsewhere. And guess what . . . she "got it." When I took the time to let Kristin know about the kind of folks flocking to the site to market to and even stalk America's youth, she instinctively understood that there are better places to hang out.

The point here is that our teens are not dumb—they just need parents to guide them. Don't let your silence about a cultural fad like MySpace be misinterpreted by your child as a stamp of approval.

I believe the Internet has the ability to be one of the great liberators of the American family. It allows parents to work from home and puts the wonders of the world at our fingertips. But it is also a web of weirdos. It's up to parents to harness the good and filter out the bad. With the millions of great sites that serve our families well, there's no need for our kids to spend time on a place like MySpace that refuses to act responsibly by filtering out the pornographers.

You wouldn't let a pervert or pedophile enter your home or speak on the phone with your child, would you? So why on Earth are you letting your kids hang out with them online?

MySpace Should Be Limited to Adults Only

Faith Alice Wright

Faith Alice Wright maintains a Web site, By Faith Only, *which contains inspirational material focused on her Christian faith.*

MySpace seems to be the big rage on the Internet. Reports say that the MySpace site is fast becoming as popular as the big boys like Yahoo and Google. This is my opinion, and my opinion only, but this popularity is not in my opinion a good thing. Parents, if you are concerned about what your children are being exposed to I suggest you keep them off MySpace.com. Sexual content, discussion of violence, foul language, drugs, alcohol, and tobacco are all discussed in an acceptable manner—it's ok to post in profiles, discussion boards, and blogs under the guise that if it doesn't harm anyone it's ok as far as I can figure. Add to that the multitude of personal photos posted that are not for some reason considered pornography but look to me much like the photos you might find on the cover of many of the adult magazines that are home delivered in plain brown wrappers.

Now, although I do not approve of such sites my time at MySpace made me believe with all my heart that this site needs to be limited to "adults only." But this is not the case! Quoted from the MySpace.com Terms of Use Agreement: "Eligibility. Use of and Membership in the MySpace Services is void where prohibited. By using the MySpace Services, you represent and warrant that (a) all registration information you submit is truthful and accurate; (b) you will maintain the accuracy of such information; (c) you are 14 years of age or older; and (d) your use of the MySpace Services does not violate any applicable law or regulation. Your profile may be deleted and your Membership may be terminated without warning, if we believe that you are under 14 years of age."

I support freedom of speech but that doesn't mean I have to have it in my house and you don't have to allow it in yours.

Yes, you read that right, 14 year old children can have memberships at MySpace, can browse all the information on the site, and can add to their "friends" list anyone else with a membership on MySpace. What does this mean? First-hand experience, a 22-year-old man bragging on his MySpace profile to be a loving father who in reality now owes one mother of one of his children over $10,000 and has not had contact with his son for 3 years and also has at least one other child whose mother decided rather than chance having this man have contact with her child, has not attempted to get child support from this man. Ok, a lot of people lie or don't tell the whole truth, but how about this MySpace member, the same "loving father", is also a known sex offender? Until recent "political debates" this person was listed publicly on the Michigan sex offender list but because of debate the list seems to only contain people who report as they are legally supposed to. If they don't report where they are then they are not listed on the current sex offender list. Yet this one man (and who knows how many others are out there) had at least five young women under the age of 18 on his friends list. Was one of them your daughter?

MySpace Does Not Verify Ages

Is it MySpace's responsibility to check the world for sex offenders? No! Is it MySpace's responsibility to check birth certificates, photo ids, or other means to verify the ages of MySpace members? No! Is it MySpace's responsibility to protect our children online? No! MySpace is online to make money, they offer free services but there is no question they are combining their Internet popularity with selling ad space to whoever wants to pay the price and those "free profiles" just keep building and building with all the trash you can imagine.

MySpace is marketed at a website offering social networking via profiles, blogs, photos, videos, blogs, and discussion groups. This mother (and grandmother) is concerned about this networking on MySpace—children under 14 merely lie about their ages and get memberships, "14" year olds are networking with sex offenders and who knows who else, and we haven't even discussed the computer problems created by the lack of control over design, profiles, and downloads.

Again, this is only one mother's opinion but MySpace needs to be not closed, but limited to adults only, and I highly recommend all parents take a minute and go to your browser "Tools", "Internet Options" and block MySpace.com—make it so your children can't even get on the site! It's much safer for your children, your computer, and your family. Leave this site for the adults who are responsible for themselves! But, PROTECT OUR CHILDREN ONLINE! I support freedom of speech but that doesn't mean I have to have it in my house and you don't have to allow it in yours.

I will add, there are most likely many good people on MySpace who aren't looking for young children, don't use foul language, don't promote violence, drugs and such. I commend those people and wish them a great experience on MySpace. My ONLY concern is the children whose parents, much like I was a few months ago, are told "it's a great site, all the kids are using it to chat online" and are not aware of what else or who else is on MySpace.com.

MySpace Exposes Teens to Crude and Disturbing Material

Catherine Saillant

Catherine Saillant is a staff writer for the Los Angeles Times.

I've covered murders, grisly accidents, airplanes falling out of the sky and, occasionally, dirty politics.

But in nearly two decades of journalism, nothing has made my insides churn like seeing what my 13-year-old daughter and her friends are up to on MySpace.com.

Here's a bulletin I recently found posted to her site:

"OMG! Add this hott guy! He will whore the first 20 people added to his friends list. . . . Add him! You can do it in his van!"

Coarse language and often profane messages . . . are plastered all over the social networking site like graffiti on bathroom walls.

Loosely translated, the teenage girl was "pimping" a teenage boy, shown smooching his guitar, as a potential new friend—or more—for my daughter. If Taylor added him to her MySpace "friends" list, the tousled-hair teen would be able to look at her website and send messages to her.

The soliciting girl made the pitch to all 245 of her own "friends" with a simple keystroke.

In the MySpace world, this is called a "whore code." It's a mild—very mild—example of the coarse language and often profane messages that are plastered all over the social networking site like graffiti on bathroom walls.

It was this coarseness and an abject lack of manners (not to mention extremely poor grammar) that bothered me the most as I entered the second month of a deal that I had worked out with my often headstrong daughter.

Though MySpace tells users that they must be at least 14 years old to join, all it takes is a casual search to see that the requirement is routinely violated. All of the kids at her junior high had MySpace accounts, Taylor pleaded. Why couldn't she?

After consulting with a circle of friends and relatives, I relented. I'd let Taylor have a MySpace site, I told her, but only if she agreed to follow some rules.

I [found] it harder and harder to balance my parental instincts with technologies that seemed to me to be rewriting the rules of adolescence.

The first was that her site would have to be set to "private." That meant that only those she had preapproved as "friends" could see her page.

Next, she could not add as a friend anyone she did not personally know.

We also agreed that no foul language or inappropriate materials could be used.

And, most important, she had to give me complete access to her site, including a password that let me view hidden e-mails.

Taylor was so excited that she immediately agreed to everything and signed the contract that we had drawn up. In the high of the moment, I felt good too. I had found a way to allow my daughter an activity that she seemed to love while protecting her from online predators—my biggest worry.

But in the days and weeks to come, our honeymoon glow would turn to alarm on my part and an increasing boldness on hers. And I would find it harder and harder to balance my parental instincts with technologies that seemed to me to be rewriting the rules of adolescence.

A Parent Discovers MySpace

This all started in late December when my cellphone rang as I was walking into a grocery store. It was the mother of one of Taylor's friends, explaining that she needed Taylor's help to shut down her own daughter's MySpace account.

Taylor, then 12, had helped the daughter set up a site without the mother's permission, and only Taylor knew the password necessary to delete it.

All of this was news to me. With an embarrassed apology, I promised to set things straight.

I didn't know much about MySpace.com then. I've since had to do my homework.

MySpace, I learned, was created by a couple of Santa Monica tech-heads, and over its two-year life, it has become the biggest website that allows people to find dates, keep in touch and socialize. If you sign onto the site, now owned by Rupert Murdoch's News Corp., you get free personal "space" to post profile information and photographs, write blogs, link music and send e-mails to other members. MySpace claims 68 million members, up more than 20 million in just the three months since I began visiting it. [As of July 2007, it had over 190 million accounts.]

Some of its fans are young adults. Many are kids like Taylor.

Our relationship had become stormy of late. Taylor resisted spending time with the family and seemed more concerned with her social life than anything else.

I wasn't surprised that she found MySpace fascinating. As a little girl, she constantly questioned me about the world around her and was not satisfied with simple answers.

She wanted to know not just how things work but why they were that way. Why are those people poor, Mommy, and why isn't everything free? If my answer didn't square with her sense of justice, there would be a whole new round of whys. She was always curious.

I could see that MySpace was a challenging and fun new universe for her to explore—only this time, she wasn't seeking my guidance.

When I confronted her about the mother's call, Taylor sheepishly admitted that she had become something of a MySpace guru for her circle of friends. She helped them set up accounts and even designed their pages if they asked.

I ... found that I was bonding with my daughter in a way that had become difficult of late.

Yes, she had her own page, Taylor said. She showed it to me.

Looking back, I realize that my reaction had little to do with the primping photographs of young girls, creepy "bulletins" and occasional foul language that I found on her site.

It was more the shock of discovering a different Taylor, a cool-teen version of the girl who used to cuddle up on the couch and watch cooking shows with me. My daughter was too young for such an uncensored world, I decided. I immediately shut her site down.

I had to ask her to do it for me. This is how she responded: "I'm really mad because it feels like you're saying I can't talk to my friends anymore. On MySpace, I get to talk to my friends and see people I don't see a lot. You get to keep in touch with everyone and it's fun. You took away my fun!"

Benefits and Drawbacks

Several things happened that made me change my mind.

My 49-year-old sister, Christine, joined MySpace and told me she was having fun using it. She urged me to set up my own account so we would have a free, easy way to exchange e-mails and photographs.

I thought about the technical skills Taylor had acquired in creating pages for herself and for her friends. And I started noticing a lot of news stories about MySpace, usually focused on its dangers.

What if I allowed Taylor to maintain a page while keeping a close eye on it? I'd join too, to become familiar with the site's benefits and drawbacks.

So I typed out the contract and Taylor enthusiastically signed it. She created an account for me and assigned me a URL, www.myspace.com/im_a_cool_mom, that reflected the giddiness of our truce.

Taylor was thrilled by this turn of events: "Yeah! . . . I can't believe you're letting me do it. I guess it's OK that I have to sign this thing. I don't want you supervising me. I think I can do it on my own. But I'm glad you let me have my own page. It's a generous thing for you to do."

Taylor was generous too. She helped me find a pretty orange background and showed me how to link to a Joni Mitchell music file that would play "Big Yellow Taxi" when a visitor landed on my Web page.

The mass mailings Taylor received each day were for me a barometer of how easily and often young teens are exposed to crude images, thoughts and words.

Meanwhile, while I was content to leave my page alone once she set it up, Taylor constantly tried out different looks on hers, as naturally as if she were changing clothes.

Almost daily she linked different music files (from a MySpace-sanctioned cache) to her site. From my daughter, I learned that Beyonce is hot and that Britney Spears definitely is not.

I also watched with surprise—and pride—as her flying fingers from memory tapped out HTML, the specialized language of Web page design, to add color and unique script to photo captions.

Using my digital camera, she took dozens of photographs of herself and her friends, posting the best to her site. Over time, I noticed how she became less interested in glamour poses, experimenting instead with shadows, light, composition and unexpected angles. She often shot photographs for her buddies' MySpace Web pages.

I was impressed and found that I was bonding with my daughter in a way that had become difficult of late.

I'd ask questions about how to perform various functions, and she'd help me out. Our technical talks usually segued into discussions of her social and academic problems and worries.

The novelty of MySpace was interesting to me. I didn't have a lot of friends, but I enjoyed chatting with the few I did.

Some Disturbing Posts

As for predators, six weeks into our experiment, Taylor had not received messages from anyone other than friends or schoolmates.

At one point a couple of teenage boys from local schools sent e-mails asking if they could come to her house after school. When she didn't respond, one of the boys sent an angry e-mail using swear words to demand an answer. With my prompting, she deleted both as friends.

Foul language was a constant, and it annoyed me. Catholic schoolgirl upbringing aside, at 45 I'm not naive or a prude. Still, the mass mailings Taylor received each day were for me a barometer of how easily and often young teens are exposed to crude images, thoughts and words.

Some of it was the type of notes my own generation passed around when the teacher wasn't looking. An ominous story, followed by the warning of misfortune: "If you don't repost this within 10 minutes, you will have bad luck for 7 years and no one will love you EVER IN YOUR LIFE!!!"

But other posts were truly disturbing, such as the cautionary, and detailed, tale of a young girl who was raped by her father, died from a sexually transmitted disease and now haunts those who read her story.

Sharing MySpace . . . did, for a brief time, provide me with a reassuring glimpse of the curious and smart girl I know so well.

Surveys are also popular and contain such telling questions as "Have you ever taken drugs?" "Are you a virgin?" and "Abortion—for or against?"

Talk about a permanent record!

I reminded Taylor that she shouldn't leave any identifying information in her writings, even if she was writing to a friend.

Once, I caught her using a swear word in a message and grounded her from MySpace for two weeks.

Another time, I asked her to delete song lyrics and other material that I considered offensive from her page. (She said she just liked the beat, but I didn't like the references to getting drunk, rape and "blowin' up the neighborhood.")

But for the most part she was being responsible. "Maybe this is going to work," I thought.

Offensive Photos

In late February, about two months after we made our pact, I logged onto Taylor's account one morning to look it over. As I scrolled down, some new photographs caught my eye.

I froze. For some reason, Taylor had posted shots of herself and two other girls giving a one-fingered salute.

I couldn't believe that she would post the photos knowing that I was looking over her shoulder. What was she thinking?

This is what she was thinking, Taylor said: "I was like, 'Should I do this? I don't know.' But I thought you wouldn't care because I didn't say anything on them. They were just pictures. And [the two other girls] already had them up on their pages."

I printed out the offending photos and confronted Taylor with them. I showed her the contract and asked her to read it. Then I told her I was shutting down her account because she had broken the rules.

She didn't take it well. She cried and pleaded for one more chance. I braced myself for the onslaught and said no.

She pleaded some more. I struggled to stay calm and resolute. After a few minutes of arguing, Taylor gave up. Her anguished reaction, before stomping into her bedroom and slamming her door, went something like this: "You've ruined my life!"

Later, we talked about what happened. Taylor knew she had blown it and was as angry at herself as she was at me. I encouraged her to join the yearbook staff at her junior high school to practice her newfound design and photography skills.

If she showed more maturity, I told her, she could reopen her MySpace account when she turned 14 next March. That minimum age requirement was making a lot of sense to me.

Taylor listened and accepted her fate. She had just one question.

"Can I get my cellphone a little earlier?"

A Reassuring Glimpse

Sharing MySpace did not magically transform my relationship with Taylor. She still circles me like an alpha dog, and we still fight like cats.

But it did, for a brief time, provide me with a reassuring glimpse of the curious and smart girl I know so well. It reminded me that once we get past the storms of adolescence, that person will still be there.

With her usual bluntness, Taylor said her lesson was simpler: "You're still a cool mom—but not THAT cool."

Or perhaps even that powerful.

In the 24 hours before her MySpace account went dormant, Taylor received one last helpful e-mail from a friend: "U know u can just make another one but have a different name," wrote the girl. "That's what I did."

The Pros Outnumber the Cons for MySpace

Melanie Galvan

Melanie Galvan maintains the blog galvanized, *which she describes as "Sparkling moments of clarity that surface between monotony and minutiae, rote and routine that are my life."*

Teens and MySpace.com—Is it a good combination? I happen to think that it is. I'm going to tell you why.

Imagine that you, a parent, schedule every teen who wants to be friends with your child for a blind interview. In that interview, you are behind a two-way mirror. Welcome to MySpace 101. This is the best way to get acquainted with your teen's friends as they are. Every person on your child's Friends List has a profile and content by which you can get to know her. Just keep one thing in mind while you read—you *must not* be judgmental.

Teens are young adults-in-progress. You may see pics on their sites that are not altogether wholesome. You may see some suggestive comments and vulgar language. You may see things on other teens' MySpace sites that even make you cringe; as a matter of fact, you most certainly will. Just remember to postpone your reactions and allow your teen to form her own opinions about what she sees. As you look over her shoulder from time to time, and her eyes roll, just ask her, "So what do you think about that?" or "Pretty cool, huh." Just keep in mind that your job is merely to monitor your own teen's MySpace site. In (eh, forcibly) sharing "the MySpace experience" with you, she is laying wide open her friends and acquaintances for your perusal, not your ridicule.

And keep in mind that this openness is strictly between your teen and you . . . and what you discuss goes no further.

Melanie Galvan, "Teen Talk: MySpace—A Good Thing?" *galvanized*, January 9, 2007. http://galvanized.wordpress.com. Reproduced by permission of the author.

The loss of coolness due to a parent's big mouth is killer in these cases. As a byproduct, this can even foster some trust along the way.

Most Teens Are Fairly Responsible

One thing that I have noticed that has been a pleasant surprise is that, for the most part, most of the kids on my teen's Friends List are fairly responsible in what they post and even maintain a good amount of propriety. A couple of them—actually, teenaged boys—have posted thoughtful and appreciable blogs with lovely original poetry and their thoughts on life, relationships, disappointments, God, and what they consider blessings. It's admirable, to me, to see teens posing questions to one another and offering insight as they discover it. A lot more exceptional teens are hanging out with my child than I would have thought before reading their sites. In their sites, I have seen great potential for becoming professional musicians, writers, photographers, web designers and future marketing professionals, models, and, yes, even intellectuals. Who knew kids could be so cerebral. (I didn't know any when I was in school ... but, then again, we didn't have MySpace where we could express ourselves from behind a screen, with some degree of anonymity.)

[MySpace can] promote closeness ... and much-needed conversations between parent and child [about] what is/is not normal, acceptable, safe, wise, kind, or honest in relationships, in society.

And let's be realistic here—there are also the partygoers and the superficial brats. But, again, remember that you would not have wanted to be judged as an adult when you were, say, 16 years old. I know that I wouldn't have. These are young people still discovering who they are, what they believe in, and where they are headed. Your job, as a parent, is to make sure that you, in a sense, keep a hand on your teen's shoulder as she navigates through this new territory that is Unbridled (well, kind of) Adolescence in All Its Glory.

Now, I realize that many of my teen's friends would cringe if they knew what I know about them. They wouldn't understand that I respect them. And while I admit that a few of their sites are pretty raunchy (this week), I'm not running to their parents with that information, and I'm not cutting them out of my child's life for being, well, pre-adults. Let's just consider MySpace a kind of *rumspringa* [a time when adolescents are allowed to experiment] for the general teen population. While I know that my child is going to venture out a bit before she leaves for college, I want her to understand that she hasn't yet left, and that I am still a part of her life. For now, while she uses MySpace, I'm still a small influence, be it invisible now, of her teenaged social life.

Also keep in mind that most parents, obviously, do not check their kids' sites. This blows me away. I know this because, if they took just a glance, they would surely insist that certain content be immediately taken off. I don't know if this is due to plain ignorance, or a lack of interest, or a shortage of time. Worse, what disturbs me is that some parents may (erroneously) be allowing their kids on this Internet playground without supervision. No, actually, it's more that Information Superhighway that we kept hearing was being built, and we've got to belt them in. Just because a child is in what appears to be an adult body does not excuse you as the parent from being . . . well . . . still the parent. To allow your teen to go it alone or, God forbid, hand-in-hand with other teens to explore the great unknown that is their published websites is just negligent. Save the stepping back for Facebook when they leave for college because, regrettably, the exposure to keggers and experimentation are inevitable . . . though your child will, hopefully, sidestep them. (That's where prayer comes in.) But until college, my teen is living at home and, therefore, will remain an open book to her dad and me, at least with respect to her MySpace and email.

The Cons for MySpace

Let's be fair, though, and listen to some of the cons, shall we?:

It's a huge waste of time and distracting from school and academia. *(Then limit the time allowed.)*

It's a playground for predators. *(Not if the profile is kept set on "private" and parents check in.)*

It's a way for kids to "hook up" and spread news of parties. *(Again, be present. Once your child knows that you know, the chain of communication is kind of broken or at least potholed, isn't it.)*

Remind your teen that whatever is posted today could come back to haunt her tomorrow, whether it be through images or written words.

It's a site for kids to publish inappropriate content and photos. *(Not your child, if you check daily. Also, your child knowing that you see her friends' sites just might cause her to suggest to her friend to take it down.)*

The Pros for MySpace

The Pros far outnumber the cons for MySpace. It:

- Gives kids an outlet to vent to peers and to see that they are normal and their struggles common.

- Allows for networking and making new friends.

- Is a great equalizer for those who are "geekier" and have skills that are otherwise not as apparent, and can even allow them to be accepted and offer help when needed.

- Is a tool for frequent validation and building self-esteem.

- Is a creative outlet for writing and self-expression. *(Teachers should be taking full advantage of this and being creative in using MySpace to reach kids.)*

- Is a terrific learning tool for web design and web presence.

- Is a site that allows kids to share art, be it through music and lyrics, poetry, photos, or video.

- Is a way for a teen to put a "spin" on his image, or create one based on how he would like to be perceived.

- Is a chance to interact with different cliques where school may not present an opportunity.

- Is a catch-all for teens contacting each other and keeping in close touch (through comments, instant messaging, or email).

- I am sure that I could think of many more . . . but I am getting tired and feel that I have pontificated enough.

But the best argument for teens on MySpace is that it's like a ready resume for parents to screen and "interview" your kid's friends. And if the parent is interested and lets her child know that her life is worthy of attention, it can even promote closeness and confidences, and much-needed conversations between parent and child that would otherwise just not come up, such as: what is/is not normal, acceptable, safe, wise, kind, or honest in relationships, in society. It's a continual opportunity for a parent to remind a teen of who she is, how valuable she is as a person, and to emphasize respect and propriety. MySpace is an excellent relationship teacher.

Above all else, your child must know that you are not judgmental of her friends and their choices. MySpace is a rewarding pasttime for a teen who just wants a presence . . . anywhere.

You may think that your teenager's life just isn't really "real" yet. Well, here's news—in case you can't remember what it was like to be a teenager—it is real to them, and should also be to you. They have the same frustrations and needs to vent as we do. They have the same drive to create and should be allowed to express themselves in writing and building an image that they feel mirrors who they are . . . or want to become. They need the chance to broaden their social circles and find their niche.

Parents Should Check Teens' Sites

So it's very easy to be the parent here. Just be there. Don't be afraid to be an imposition to your child or to even anger them. Make sure that your child's site is set to "private" if he is under age 18. Screen his site from time to time. Make sure it's free of profanity and vulgarity. Have him delete offensive comments. Remind your teen that whatever is posted today could come back to haunt her tomorrow, whether it be through images or written words. (Regrettably, many kids don't have the capacity to consider consequences that could arise someday in submitting a college application or interviewing for a job.) Make sure that your teen's site—profile content and photos—are appropriate. Ensure that no personal information is listed that could leave them open to predators. Watch to see what new friends appear, as it could show you where your child is headed. Ask your kid questions. It's just that easy.

And here is the doozie—the Mother of All Impositions of Parent on the Child, sure to produce animosity and, in some cases, extreme loathing—know all of your child's passwords, especially MySpace password and email address passwords. Be open and honest with them that, while it is their own personal site, you will randomly check it. Explain to your child that, in the event that anything is inaccessible to you, her MySpace site will be shut down and the computer off limits. It's a trust issue—but the parent has to be realistic in her expectations and not abuse the situation.

Above all else, remember that your child is not an angel, but a teenager. Allow them some ample roaming distance and some chances to grow up and make their own judgments. However, set clear boundaries, lines that must not be crossed—for example, with regard to romantic relationships, alcohol/drug use, driving, sneaking out, meanness, and bullying.

To make it short, the development of your child's social life is going to happen whether you're there or not. First, be the parent and, second, the confidante. It comes down to three simple words—"Give a damn." There are too many parents out there who just don't.

Social Networking Plays an Important Role in Young People's Lives

Gary Stager

Gary Stager is senior editor for District Administration *magazine and an adjunct professor at Pepperdine University.*

During seventh grade a friend and I created a publication as an alternative to the school newspaper. It was quite a challenge in the days before access to photocopiers, but entertaining our handful of readers made the effort worthwhile.

I remember the day when the faculty advisor of the official school newspaper followed me into the boy's room, threw my 50-pound body against the wall and threatened to kill me if we published another issue. The English faculty's Tony Soprano really schooled me in the subtleties of the First Amendment. Ah, life was so much simpler then.

Since man first scribbled on cave walls and peed in the snow, humans have been compelled to share their stories. Recent decades have seen great violence done to student expression through court-sanctioned censorship of student publications and other forms of adult supremacy. Issues of critical importance and interest to students are banned from student newspapers and classroom discussions. Political correctness and tolerance are used to masquerade for intolerant policies like "zero-tolerance" and increasingly mediocre curriculum. High school credit is awarded just to get kids to contribute to some school newspapers.

Back in the good old days of lavatory justice, children climbed trees, played ball in the street and joined the scouts

Gary Stager, "Guess Why They Call It MySpace," *District Administration*, vol. 42, May 2006, pp. 78–79. Copyright 2006 Professional Media Group LLC. Reproduced by permission.

so they could play with fire. Santa delivered chemistry sets complete with recipes for gunpowder and kids could get together without having "my people call your people." You could actually read all 30 pages of *Sarah Plain and Tall* without a textbook publisher excerpting it for you. Remember when you could read a book without being interrupted every paragraph to answer a comprehension question?

Same Problems, Modern Times

Parents and educators have done a lot more to wreck childhood than [inventor of the World Wide Web] Tim Berners-Lee (ask a kid to show you how to learn about him at Wikipedia). Schools endanger the very students they seek to protect when they bubble-wrap kids and the curriculum. School principals are banning classic plays like *Grease* and *The Crucible* while childish schlock like *Seussical* is now the most performed high school musical. John Taylor-Gatto argues that the mission of schools is now to extend childishness through graduation. Dependency and fear retard the learning process. It is difficult, if not impossible, for students to develop moral values and solve ethical dilemmas when school never allows them to make a decision or mistake.

Much of MySpace's content is inane, but we should avoid destroying the place 21st century kids built for themselves.

Every generation has had to wrestle with understanding new media. In 1954, the U.S. Senate held hearings to investigate how comic books harm children. Who can forget Tipper Gore vs. Frank Zappa [who testified as opponents at a senate hearing on porn rock] or the 1995 *Time* magazine cover depicting a computer-induced zombie child with CyberPorn in block letters? The educational technology community has a similar level of paranoia manifest in discussions over whether students should have their own floppy, be allowed to save on the hard drive, surf the Web, send an e-mail or use a USB key. It is impossible to discern the lines between genuine safety concerns and tyranny.

The latest episode of adults behaving badly involves the hysteria over the popular Web site, MySpace. MySpace is a social networking site where anyone can publish and maintain relationships with friends. Chances are that your only experience with MySpace has come through local TV news stories about how parents must rescue their teenagers from this deadly cyber-sewer before sports and weather. It's a fair bet that you are not one of MySpace's 66 million registered users [over 190 million as of July 2007]. There has probably never been a more aptly named product. They call it MySpace because it belongs to them, not you.

MySpace provides users with Web space where they may share their thoughts and creative output with classmates and friends around the world. What makes sites like MySpace different from other blogging sites is that you may ask interesting people to be your friend. Then you'll know when your friends are online, who their friends are and quickly develop affinity groups. You may organize communities around interests, geography or a host of other variables. You can chat via instant messages, insert music in your page and share all the photos and doodads that kids use to decorate their locker. (If school still trusts them with lockers.)

The recent student walkouts over the proposed immigration bill were organized on MySpace. The role such sites play in grassroots and electoral politics is inestimable.

In fact; more than one observer has compared MySpace pages to a teenager's bedroom walls. My 12th grade daughter's MySpace site is unbearable. Animated gifs, flashing graphics, dopey poses, horrific music, yellow text on hot pink backgrounds and other elements of Web design hell assault your senses until you run away or quit the browser. Much of MySpace's content is inane, but we should avoid destroying the place 21st century kids built for themselves.

Risks of MySpace Are Avoidable

Sure there are creeps using MySpace. That's why you need to teach children not to share personal information online or get in a car with strangers. MySpace never shows the real name of a member, just a pseudonym like a CB radio handle. When you add a person to your friends list, that person receives an e-mail asking for permission. If someone turns out to be unpleasant, you may ban him or her from contacting you with a click. Even critics of MySpace concede the company is incredibly responsive to concerns over online ickyness. A student may be at greater risk of being suspended by her school for something written at home on MySpace as there is of that teenager being physically harmed.

MySpace is changing how young people communicate, collaborate and spend their discretionary funds. Network TV programs are being launched on MySpace and countless bands have experienced enormous sales due to word-of-mouth and users sharing music with their friends. The recent student walkouts over the proposed immigration bill were organized on MySpace. The role such sites play in grassroots and electoral politics is inestimable.

Today my daughter's high school experienced a small fire. She learned of the fire from friends via MySpace nearly 11 hours before the local television news reported it. MySpace is a teenager's record store, newsstand, community center, fan club and 24/7 news network. As our government strives to spread democracy abroad, we would be well served by celebrating the electronic democracy afforded by sites like MySpace.

I just learned that my daughter has retaliated for me showing her MySpace site in conference presentations by posting an unflattering photo of me online. I wonder if I can get the local school district to punish her?

Teens Need Privacy from Parents

Michael Thompson

Michael Thompson is a psychologist, school consultant, and author or coauthor of seven books.

To pry or not to pry? That is the question. As a child psychologist who consults to schools, I am constantly asked questions about what goes on in the minds of children. Parents in particular seem to think that it is always a good thing to know everything about what their kids think and what they do. Recently, I received a phone call from a middle-school principal who told me that the parents of his students had become concerned about their children's online activities. They were clueless about what their kids were doing online, and too naive to do anything about it. They told him that they feared the popular social-networking Web site MySpace and they worried that their kids could be victims of online bullying, or worse. The principal asked if I would visit the school to address these concerns, and I agreed to go.

Just minutes later, I received an e-mail from the head of an elementary school who took another point of view. "When we were growing up," he wrote, "Most of us had a lot of time on our own that we as children filled on our own. I am wondering if one of the reasons that kids spend a great deal of time instant messaging and making websites on MySpace is to create a sense of privacy or a world apart from their parents. We tend to know what the negatives of instant messaging and personal websites might be, but what might the positives be? Maybe we are missing something that we need to keep in mind. Do they need privacy from their parents and, if so, why do they?"

Children Need Privacy

My answer was simple: yes, children do need privacy. How do you know who you are and what you can really do unless you actually have a chance to be on your own? The most treasured memories of my childhood are walking around the streets of New York City accompanied by my friends, with that precious ticket to freedom—the bus pass—in my pocket.

Other sweet memories were from July and August, when I spent hours away from my parents near a lake in southern Massachusetts accompanied by friends and cousins. We played for hours; we built forts and fought imaginary enemies. Did we do some bad things? Sure, I guess so. My friends and I blew up some bullfrogs with firecrackers. We tried smoking cigarettes; we made some illicit campfires. We even talked some girls into playing "doctor." All of this before I was 11 years old.

Would my parents have been upset if they had known what we were doing? Without a doubt. Did we take some risks? Certainly. I don't know whether I'd be better or worse off had I taken less risks as a child, but I do know that I cannot imagine my childhood without those times. I cannot conceive of my adult personality without those memories.

Kids today spend almost no time "in the woods," and their moments spent away from their parents' watchful gaze are precious and few. In our middle and upper-middle-class neighborhoods, children are largely indoors, taking lessons doing homework and getting ready to go to town sports. They are endlessly supervised and monitored. How many parents today are willing to do what our parents did: shovel us out the door at noon, saying "Don't come back until 6 o'clock?"

The Internet Is the Only Place Children Go Alone

We cannot bear the anxiety of not knowing everything about our children's whereabouts in the physical world. So our children wander off into cyberspace, killing zombies, talking to strangers and visiting all kinds of Web sites, weird and stupid and sexy. And they know we don't have the time, attention or expertise to follow them there. The Internet is often the only private place for a child today.

That isn't to say that parents and administrators should turn a blind eye to their children's online activities, they shouldn't. To the middle-school principal, I suggested he ask his computer teacher to go online and check out the MySpace pages of all of his students. After all, a middle-school director in Maryland I worked with found that two of his girls had posted photos of themselves in their underwear, and their parents knew nothing about it. We need to be vigilant.

But I'm always torn when people call me for wise counsel about kids and privacy, because roaming around in my head there is still a child who treasures his private adventures. My inner boy is certainly going to shout, "Let them be! Let them take some risks." The parent in me is going to worry and advise, "Check their Web sites. There are dangers out there. There are pedophiles on the Internet." And what about the psychologist in me?

I hope my inner psychologist has the courage to remember his own boyhood and to keep reminding parents of how precious a bit of privacy was to them when they were growing up.

Teens Need a Place to See and Be Seen by Peers

danah boyd

danah boyd is a PhD candidate at the School of Information at the University of California-Berkeley and a fellow at the University of Southern California Annenberg Center for Communications.

I want to talk with you today about how teenagers are using a website called MySpace.com. I will briefly describe the site and then discuss how youth use it for identity production and socialization in contemporary American society.

I have been following MySpace since its launch in 2003. Initially, it was the home to 20-somethings interested in indie music in Los Angeles. Today, you will be hard pressed to find an American teenager who does not know about the site, regardless of whether or not they participate. Over 50 million accounts [over 160 million as of March 2007] have been created and the majority of participants are what would be labeled youth ages 14–24. MySpace has more pageviews per day than any site on the Web except Yahoo! (yes, more than Google or MSN).

Many of you may have heard about MySpace, most likely due to moral panic brought on by the media's coverage of the potential predators and bullying. There is no doubt that the potential is there, but there are more articles on predators on MySpace than there have been reported predators online. Furthermore, bullying is a practice that capitalizes on any available medium. Moral panics are a common reaction to teenagers when they engage in practices not understood by adult culture. There were moral panics over rock & roll, television,

danah boyd, "Speech: Identity Production in a Networked Culture: Why Youth (Heart) MySpace," American Association for the Advancement of Science, February 19, 2006. Reproduced by permission of the author.

jazz and even reading novels in the early 1800s. The media, typically run by the parent generation, capitalize on and spread the fear with little regard for data or actual implications. Examples are made out of delinquent youth, showing how the object of fear ruined them in some way or other. The message is clear—if you don't protect your kids from this evil, they too will suffer great harm to their minds, bodies or morals.

There *are* potential risks on MySpace but it is important not to exaggerate them. The risks are not why youth are flocking to the site. To them, the benefits for socialization outweigh the potential harm. For this reason I want to ask you to put your fears aside for the duration of this presentation and try to see the values of MySpace for youth. What they are doing is really fascinating.

What Is MySpace?

So what is MySpace? MySpace is a social network site. In structure, MySpace is not particularly unique. The site is a hodgepodge of features previously surfaced by sites like Friendster, Hot or Not, Xanga, Rate My Teacher, etc. At the core are profiles that are connected by links to friends on the system. Profiles are personalized to express an individual's interests and tastes, thoughts of the day and values. Music, photos and video help users make their profile more appealing.

> For most teens, [MySpace] is simply a part of everyday life—they are there because their friends are there and they are there to hang out with those friends.

The friend network allows people to link to their friends and people can traverse the network through these profiles. An individual's "Top 8" friends are displayed on the front page of their profile; all of the rest appear on a separate page. Bands, movie stars, and other media creators have profiles within the system and fans can friend them as well. People can comment on each others' profiles or photos and these are typically displayed publicly.

Originally, the site was 18+ and all data was public. Over time, the age limit dropped to 16 and then, later, to 14. The youngest users are given the option to make their profiles visible to friends-only and they do not appear in searches.

When someone starts an account, they are given an initial friend—Tom Anderson, one of the founders of MySpace. By surfing the site, they find and add additional friends. Once on MySpace, most time is spent modifying one's own profile, uploading photos, sending messages, checking out friends' profiles and commenting on them. Checking messages and getting comments is what brings people back to MySpace every day.

When MySpace was initially introduced, skeptics thought that it would be just another fad because previous sites like Friendster had risen and crashed. Unlike the 20-somethings who invaded Friendster, the teens have more reason to participate in profile creation and public commentary. Furthermore, MySpace's messaging is better suited for youths' asynchronous messaging needs. They can send messages directly from friends' profiles and check whether or not their friends have logged in and received their email. Unlike adults, youth are not invested in email; their primary peer-to-peer communication occurs synchronously over IM. Their use of MySpace is complementing that practice.

Many teens access MySpace at least once a day or whenever computer access is possible. Teens that have a computer at home keep MySpace opened while they are doing homework or talking on instant messenger. In schools where it is not banned or blocked, teens check MySpace during passing period, lunch, study hall and before/after school. This is particularly important for teens who don't have computer access at home. For most teens, it is simply a part of everyday life—

they are there because their friends are there and they are there to hang out with those friends. Of course, its ubiquitousness does not mean that everyone thinks that it is cool. Many teens complain that the site is lame, noting that they have better things to do. Yet, even those teens have an account which they check regularly because it's the only way to keep up with the Joneses.

Of course, not all teens are using the site, either because they refuse to participate in the teen fad or because they have been banned from participating. Such non-conformity is typical of all teen practices.

Most teens are concerned with resolving how they perceive themselves with how they are perceived.

With this framework in mind, I want to address three issues related to MySpace: identity production, hanging out and digital publics.

Profiles

Every day, we dress ourselves in a set of clothes that conveys something about our identity—what we do for a living, how we fit into the socio-economic class hierarchy, what our interests are, etc. This is identity production. Around middle school, American teens begin actively engaging in identity production as they turn from their parents to their peers as their primary influencers and group dynamics take hold.

Youth look to older teens and the media to get cues about what to wear, how to act, and what's cool. Most teens are concerned with resolving how they perceive themselves with how they are perceived. To learn this requires trying out different performances, receiving feedback from peers and figuring out how to modify fashion, body posture and language to better give off the intended impression. These practices are critical to socialization, particularly for youth beginning to engage with the broader social world.

Because the teenage years are a liminal period between childhood and adulthood, teens are often waffling between those identities, misbehaving like kids while trying to show their maturity in order to gain rights. Participating in distinctly adult practices is part of exploring growing up. Both adults and the media remind us that vices like sexual interactions, smoking and drinking are meant for adults only, only making them more appealing. More importantly, through age restrictions, our culture signals that being associated with these vices is equal to maturity.

The dynamics of identity production play out visibly on MySpace. Profiles are digital bodies, public displays of identity where people can explore impression management. Because the digital world requires people to write themselves into being, profiles provide an opportunity to craft the intended expression through language, imagery and media. Explicit reactions to their online presence offers valuable feedback. The goal is to look cool and receive peer validation. Of course, because imagery can be staged, it is often difficult to tell if photos are a representation of behaviors or a representation of them.

By going virtual, digital technologies allow youth to (re)create private and public youth space while physically in controlled spaces.

Comments

On MySpace, comments provide a channel for feedback and not surprisingly, teens relish comments. Of course, getting a comment is not such a haphazard affair. Friends are *expected* to comment as a sign of their affection. Furthermore, a comment to a friend's profile or photo is intended to be reciprocated. It is also not uncommon to hear teens request comments from each other in other social settings or on the bulletin boards. In MySpace, comments are a form of cultural currency.

For those seeking attention, writing comments and being visible on popular people's pages is very important and this can be a motivation to comment on others' profiles. Of course, profile owners have the ability to reject comments and Tom rejects most of them. Some people literally spam their network with comments. Last week, there were "Valentine's cards" that people made and added to the profiles of all of their friends via comments. People advertise events through mass comments. Some comments are also meant to be passed on, creating virus like memes.

The rules of friending are also very important. It is important to be connected to all of your friends, your idols and the people you respect. Attention-seekers and musicians often seek to be friended by as many people as possible, but most people are concerned with only those that they know or think are cool. Of course, a link does not necessarily mean a relationship or even an interest in getting to know the person. "Thanks for the add" is a common comment that people write in reaction to being friended by interesting people.

While these dynamics may not seem particularly important, they are essential to youth because they are rooted in the ways in which youth jockey for social status and deal with popularity. Adults often dismiss the significance of popularity dynamics because, looking back, it seems unimportant. Yet, it is how we all learned the rules of social life, how we learned about status, respect, gossip and trust. Status games teach us this.

Hanging Out

So what exactly are teens *doing* on MySpace? Simple: they're hanging out. Of course, ask any teen what they're *doing* with their friends in general; they'll most likely shrug their shoulders and respond nonchalantly with "just hanging out." Although adults often perceive hanging out to be wasted time, it is how youth get socialized into peer groups. Hanging out amongst friends allows teens to build relationships and stay connected. Much of what is shared between youth is culture—fashion, music, media. The rest is simply presence. This is important in the development of a social worldview.

MySpace allows youth to interact with [a] broader peer group rather than simply being fed information about them from the media.

For many teens, hanging out has moved online. Teens chat on IM for hours, mostly keeping each other company and sharing entertaining cultural tidbits from the web and thoughts of the day. The same is true on MySpace, only in a much more public way. MySpace is both the location of hanging out and the cultural glue itself. MySpace and IM have become critical tools for teens to maintain [in the Words of Misa Matsuda] "full-time always-on intimate communities" where they keep their friends close even when they're physically separated. Such ongoing intimacy and shared cultural context allows youth to solidify their social groups.

Adults often worry about the amount of time that youth spend online, arguing that the digital does not replace the physical. Most teens would agree. It is not the technology that encourages youth to spend time online—it's the lack of mobility and access to youth space where they can hang out uninterrupted.

Classes of Space

In this context, there are three important classes of space: public, private and controlled. For adults, the home is the private sphere where they relax amidst family and close friends. The public sphere is the world amongst strangers and people of all statuses where one must put forward one's best face. For most adults, work is a controlled space where bosses dictate the norms and acceptable behavior.

Teenager's space segmentation is slightly different. Most of their space is controlled space. Adults with authority control the home, the school, and most activity spaces. Teens are told where to be, what to do and how to do it. Because teens feel a lack of control at home, many don't see it as their private space.

To them, private space is youth space and it is primarily found in the interstices of controlled space. These are the places where youth gather to hang out amongst friends and make public or controlled spaces their own. Bedrooms with closed doors, for example.

Adult public spaces are typically controlled spaces for teens. Their public space is where peers gather en masse; this is where presentation of self really matters. It may be viewable to adults, but it is really peers that matter.

Teens have increasingly less access to public space. Classic 1950s hang out locations like the roller rink and burger joint are disappearing while malls and 7/11s are banning teens unaccompanied by parents. Hanging out around the neighborhood or in the woods has been deemed unsafe for fear of predators, drug dealers and abductors. Teens who go home after school while their parents are still working are expected to stay home and teens are mostly allowed to only gather at friends' homes when their parents are present.

Restricting youth to controlled spaces typically results in rebellion and the destruction of trust.

Additionally, structured activities in controlled spaces are on the rise. After school activities, sports, and jobs are typical across all socio-economic classes and many teens are in controlled spaces from dawn till dusk. They are running ragged without any time to simply chill amongst friends.

Digital Publics

By going virtual, digital technologies allow youth to (re)create private and public youth space while physically in controlled spaces. IM serves as a private space while MySpace provide a public component. Online, youth can build the environments that support youth socialization.

Of course, digital publics are fundamentally different than physical ones. First, they introduce a much broader group of peers. While radio and mass media did this decades ago, MySpace allows youth to interact with this broader peer group rather than simply being fed information about them from the media. This is highly beneficial for marginalized youth, but its effect on mainstream youth is unknown.

The bigger challenge is that, online, youth publics mix with adult publics. While youth are influenced by the media's version of 20somethings, they rarely have an opportunity to engage with them directly. Just as teens are hanging out on MySpace, scenesters, porn divas and creatures of the night are using MySpace to gather and socialize in the way that 20somethings do. They see the space as theirs and are not imagining that their acts are consumed by teens; they are certainly not targeted at youth. Of course, there *are* adults who want to approach teens and MySpace allows them to access youth communities without being visible, much to the chagrin of parents. Likewise, there are teens who seek the attentions of adults, for both positive and problematic reasons.

That said, the majority of adults and teens have no desire to mix and mingle outside of their generation, but digital publics slam both together. In response, most teens just ignore the adults, focusing only on the people they know or who they think are cool. When I asked one teen about requests from strange men, she just shrugged. "We just delete them," she said without much concern. "Some people are just creepy." The scantily clad performances intended to attract fellow 16-year-olds are not meant for the older men. Likewise, the drunken representations meant to look "cool" are not meant for the principal. Yet, both of these exist in high numbers online because youth are exploring identity formation. Having to simultaneously negotiate youth culture and adult surveillance is not desirable to most youth, but their response is typically to ignore the issue.

> *What we're seeing right now is a cultural shift due to the introduction of a new medium and the emergence of greater restrictions on youth mobility and access.*

Parents also worry about the persistence of digital publics. Most adults have learned that the mistakes of one's past may reappear in the present, but this is culturally acquired knowledge that often comes through mistakes. Most youth do not envision potential future interactions.

A Place for Peers

Without impetus, teens rarely choose to go private on MySpace and certainly not for fear of predators or future employers. They want to be visible to other teens, not just the people they they've friended. They would just prefer the adults go away. All adults. Parents, teachers, creepy men.

While the potential predator or future employer don't concern most teens, parents and teachers do. Reacting to increasing adult surveillance, many teens are turning their profiles private or creating separate accounts under fake names. In response, many parents are demanding complete control over teens' digital behaviors. This dynamic often destroys the most important value in the child/parent relationship: trust.

Youth are not creating digital publics to scare parents—they are doing so because they need youth space, a place to gather and see and be seen by peers. Publics are critical to the coming-of-age narrative because they provide the framework for building cultural knowledge. Restricting youth to controlled spaces typically results in rebellion and the destruction of trust. Of course, for a parent, letting go and allowing youth to navigate risks is terrifying. Unfortunately, it's necessary for youth to mature.

What we're seeing right now is a cultural shift due to the introduction of a new medium and the emergence of greater restrictions on youth mobility and access. The long-term implications of this are unclear. Regardless of what will come, youth are doing what they've always done—repurposing new mediums in order to learn about social culture.

Technology will have an effect because the underlying architecture and the opportunities afforded are fundamentally different. But youth will continue to work out identity issues, hang out and create spaces that are their own, regardless of what technologies are available.

Adults Should Become Role Models for Teens on MySpace

Stephen Downes

Stephen Downes is a researcher at the Institute for Information Technology's Internet Logic Research Group in New Brunswick, Canada. He maintains a Web site, Stephen's Web, focused on innovation in the use of online media in education.

There is a large community that believes that sites and services such as MySpace are genuinely dangerous and that it is irresponsible to allow children to simply wander through them at will.

As one person writes to me from Montana, for example, "they are the online equivalent of kids attending a cocktail party and mixing with adult strangers of every shape and motivation."

I appreciate and share this concern for the safety of children. But my reaction is tempered by what I would consider to be a sense of proportion. Specifically, when presented with something like MySpace, I ask, first, how dangerous it is, and second, whether the proposed measure makes sense given the level of danger.

The concern is that unfettered and unrestrained access to the adult world exposes children to ideas and behaviours that are disturbing.

The concern here, I think, isn't that some stranger will use a MySpace account to hunt down a child. There's no evidence that I have seen showing that a child is in more danger online than, say, in the home. Parents and friends continue to be the major abusers of children. MySpace hasn't changed that.

Stephen Downes, "Adults and MySpace," *Half an Hour*, June 15, 2006. http://halfanhour.blogspot.com. Reproduced by permisson of the author.

The concern is that unfettered and unrestrained access to the adult world exposes children to ideas and behaviours that are disturbing, ideas and behaviours that children are not yet ready to comprehend, much less emulate as children do. And the concern is that parents and teachers do not know enough about the Internet to understand the impact of prolonged exposure to the adult world online.

Fair enough. I think this is a good point. We know that media has an impact on children. That's why advertisers advertise; they would not spend the hundreds of millions of dollars if such advertisements did not change attitudes and (hence) increase sales. And we know, therefore, that what children see on the Internet will change their attitudes.

Some of this impact has been documented already through the work of people like [educational software developer] Marc Prensky and others studying the 'digital generation'. It is probably too early to draw a causal connection between new media and the attitudes of today's young, since so many other influences (such as old media) are still at play. Nonetheless, when we see that children seem to be constantly online and connected to their friends, that they explore and take chances on their own, and that they expect the answers to always be in Google, we can point to the Internet and say it is certainly having some effect.

But is it a negative effect? The writer from Montana points out, "I fear that we are raising a generation that will be unable to manually sift through text to determine arguments, core points or concepts."

I fear this too, perhaps more acutely, for as an expert in logic I am constantly concerned about the basic errors in reasoning and criticism I see every day around me. My own observation, though, is that people who spend more time online are more able to deal with these issues, that the people who accept things uncritically—especially when presented from an authoritative source—are those who are uneducated and those who are mostly offline.

In other words, what I am saying is that my own observation suggests that prolonged exposure to the Internet makes someone more able to reason critically, not less. And in my more cynical days I suggest that it is this increased capacity to reason that is sparking concern among some adults.

There is a lot of bad behaviour on the Internet, and even the briefest observation shows that it is the adults, and not the kids, who are behaving badly.

Because, frankly, it seems to me that a lot of the things kids (and adults!) are told are dangerous simply aren't so, and are reflective more of a prevailing morality than of a concern for the well being of a child. I do not see, for example, how the mere presentation of naked humans is dangerous to a child, and I am indeed much more concerned about the images of abuse and violence broadcast on television in the news and in shows like C.S.I.

Nor am I concerned about things like discussions of drugs and politics and religion—it seems to me healthy for children to be exposed to the multiple views (and the more than occasional hypocrisy) that surround such issues. Being presented a wide spectrum of opinion, rather than a single point of view, teaches someone very quickly to draw their own conclusion, and to not depend on someone else (no, not even parent or teacher) for the right answer.

This is all reasonable, but I think the point of the 'cocktail party' example is that children online are presented not merely with adult behaviour, but with adults behaving badly.

"Look at who is modeling online behavior in myspace—it is largely adults who post provocative photos and language. Kids see this, take it as acceptable and cool, and do it themselves. Again, typical behavior for teens. Add a lack of modeling of positive or acceptable behavior in this venue by teachers and parents, and voila—negative models yield negative results. How much time do you spend posting things on my-space? I don't. Who does?—people with an axe to grind, show offs, exhibitionists, etc."

I think there is no question that there is a lot of bad be-haviour on the Internet, and even the briefest observation shows that it is the adults, and not the kids, who are behaving badly. And in spaces such as MySpace, it does seem that the only adult presence is a negative one.

Is our best response, though, to kick the kids off MySpace? My first reaction seems to be that we are punishing the kids for the actions of the badly behaved adults.

After all, if a grown man came to a school playground and started swearing and drinking and making lewd remarks, we would react by removing the adult, not by preventing children from accessing the park.

Children, if they understand anything, understand justice. . . . They know that punishing an entire age group for the actions of a few is unfair.

The point is, it is up to adults to moderate the behaviour of adults. And if children are not being presented proper role models, then it is up to adults to ensure that such models are available to them. And the way to do that is not to shield them from all possible role models, because that negates the benefit of the Internet. The way to do it is to be present in this space, to moderate the adults who are behaving badly, and to ourselves act as reasonable role models.

I wrote the other day, of the bullying and the cheating and the other bad behaviours that kids engage in online, "it seems to me that the problem isn't MySpace—the problem is the school. I mean, why not look at these behavours—now that they are public—and ask why students engage in them. In-stead of trying to hide everything again by blocking MySpace, or to punish people after the fact, why not ask, 'what would lead a student to think that this is appropriate?'"

And while I didn't answer that question, to me the answer is pretty evident: they think such behaviours are appropriate because they see instances of such behaviours all around them, and not the least by the authority figures governing their lives, by their teachers, their parents and their elected officials.

The children act badly, not because they are exposed to MySpace, but because, in all aspects of life, we show them with our own actions that such behaviour is acceptable.

And—most importantly for me—when we do things like ban them from sites like MySpace, we are continuing to show them that such behaviour is acceptable. Because we conjure up dangers that simply aren't there, because we act arbitrarily and with threat of force rather than with reason, and because we then hypocritically engage in exactly the behaviours we are trying to teach children not to emulate.

Children, if they understand anything, understand justice, which is why it is essential to employ precision calipers when dividing the chocolate cake among them. They know that punishing an entire age group for the actions of a few is unfair. They know that prohibiting an action, or a web site, just in case they act badly some time in the future, is unfair.

It seems to me that spaces such as MySpace are less dangerous to kids than the family television or the neighborhood park.

Finally, children have a sense of proportion. They know that being on MySpace is importantly not like being at a cocktail party because they can hit 'disconnect' at any time. When in the actual presence of adults, kids are powerless—they cannot escape and they cannot fight back. And sometimes those adults they must trust the most turn on them.

Unless we lock kids in their rooms all day, kids will interact with adults. Far better that they interact with them, and learn of the dangers, online rather than in person. Far better that their first encounter with a child molester is in a chat room rather than a city park or a movie theatre. Far better they interact in an enviroment that can be monitored and watched very easily by parents and police.

When we think of it—and we should—the rise of sites like MySpace is a boon to parents and teachers. After all, so far as I can judge, kids have always pushed the limits of authority (I know I certainly did). But now they are doing it in a public space, where we can see what they are doing, and if we can restrain ourselves a bit, and give the kids some room to grow, we can watch out for the really dangerous things, and act before anything really bad happens.

Things like bullying, for example, become much more evident. A child being abused may be able to express this in various ways online. Suicidal trends may appear on a website far sooner than in the home or classroom. Excessive use of drugs or alcohol will be reflected online and hidden in the home. And so much more! Why would we block this, the child's best hope of letting us know that something is wrong?

It is, as I said, about the level of danger, and about the proportionality of the response. And it seems to me that spaces such as MySpace are less dangerous to kids than the family television or the neighborhood park, and it seems to me that banning kids from MySpace altogether is, in that light, an excessive response.

And, indeed, if there is any action I would recommend that we as adults take, it would be to behave as adults. That, more than any other action we take, will have the most profound impact on our children. Though, I admit, asking adults to act like adults may be more idealistic than practical.

Chapter 2

Should Congress Require Schools and Libraries to Block Networking Sites?

Chapter Preface

In January of 2006 and again in April, the TV news program *Dateline* aired reports about the social networking site MySpace, with which relatively few adults were familiar at the time. "Parents are clueless," said Parry Aftab, an Internet safety expert consulted on the April show. "They're caught like deer in the headlights." There had been several recent cases involving sexual predators who had assaulted young girls they had contacted through MySpace. There was justified concern about the unsafe posting of personal information there by teens. A wave of fear-inspiring media coverage followed.

According to a March 2006, article in the *New York Times*, some experts, of whom Aftab was one, were saying "that a fear of networking sites has grown disproportionately to actual demonstrated threats, and that there is an unjustified paranoia about the sites." But the public was, in Aftab's words, "freaked." Furthermore, when parents began visiting MySpace, they saw more immediate cause for dismay than the possibility of contact by predators. Like the *Dateline* reporter, who "found scenes of binge drinking, apparent drug use, teens posing in underwear, other members simulating sex, and in some cases even having it," they encountered pages with inappropriate content, which led them to conclude that the entire site was undesirable.

The point of most media coverage was to warn parents that their children should be supervised at social networking sites and should be taught safety rules. Some people, however, felt that the government ought to take action to protect children. Many of them contacted their representatives in Congress. As a result, a bill titled the Deleting Online Predators Act (DOPA) was introduced in the House of Representatives. This law would have required schools and libraries that accept federal funds to block access to social networking sites.

The bill was written hurriedly. One major problem with it was that its wording was so broad that it would have required blocking of all commercial Web sites that allow posting of personal information and communication between users— including Amazon, eBay, and countless forums, as well as some sites that are specifically educational. Even its supporters admitted that it was not perfect, but they believed it was better than nothing.

Opponents, on the other hand, objected to the principle of the bill. Many argued that it would not make the Internet safer, since teens who could not access social networking sites in schools or libraries would simply do it somewhere else— whereas schools and libraries were the very places where they should be taught to use such sites safely. Others pointed out that children from poor and minority families who do not have Internet access at home would be hurt the worst by being deprived of the beneficial opportunities such sites offer. Still others said that decisions about blocking sites should be made locally rather than by the federal government.

DOPA was strongly opposed by the American Library Association, the National School Boards Association, and the National PTA, among other organizations. Nevertheless, to the surprise of observers who had assumed it was favored only by conservatives, the bill passed in the House of Representatives by a vote of 410 to 15. Many felt that because it was an election year, members of Congress had not wanted voters to think that they didn't care about protecting children.

A bill does not become law unless it is passed by both the House of Representatives and the Senate. DOPA was not considered by the Senate before the end of the 2006 session, which means that it was allowed to die. However, a similar bill was introduced in the Senate in January of 2007. Whether it will be passed remains to be seen.

Safeguards Are Needed at Schools and Libraries

Greg Abbott

Greg Abbott is the attorney general of the state of Texas.

M r. Chairman and members of the subcommittee, my name is Greg Abbott, I am the Attorney General for the state of Texas, and I thank you for the opportunity to testify before you today.

Let me start by thanking you for shining a spotlight on the growing national nightmare of the Internet being used as a playground for child predators. Your legislative proposal—appropriately named "Deleting Online Predators Act"—is an important step towards making the Internet safer for our children and families. I wish the solution to the growing problem was as easy as hitting a delete button. Unfortunately, it is much more complex. With your active involvement, though, Internet safety will become more of a reality for millions of American children.

The dangers to children created by social networking websites and chat rooms are very real. The Texas experience is both illustrative—and alarming.

Three years ago, we created a Cyber Crimes Unit in the Office of the Attorney General of Texas. One of its primary missions was to find, arrest and convict child predators who use the Internet to stalk their prey. The unit's nationally recognized success is tragic evidence of the risk children face when using chat rooms or social networking websites.

Our Cyber Crimes Unit has investigators who log onto chat rooms that are used by teenagers. Sometimes they log on to social networking sites like MySpace. The investigators typi-

Greg Abbott, testimony before the U.S. House Energy and Commerce Committee, Subcommittee on Telecommunications and the Internet, July 11, 2006. Reproduced by permission of the author.

cally assume the identity of a teenage girl, usually around the age of 13 or 14. Not long after they log on and assume an under-age identity, they are barraged with aggressive and vulgar language that is uninvited. All too often, the offensive Internet "chat" turns into action. The predator sets a time, date and location to sexually assault what he believes to be a 13 or 14 year old girl. On more than 80 occasions, the predator has shown up at the location of his choosing to act out on his criminal intent. The location is sometimes a motel, sometimes an apartment, sometimes a parking lot, sometimes other places. It is not uncommon for the predator to bring things like condoms, alcohol, even a bed. Each of those 80+ occasions has resulted in an arrest of the child predator by the Texas Attorney General's office.

As an example, we recently arrested a 50-year-old man after he showed up at a Central Texas restaurant to meet what he thought was a 14-year-old girl. This predator had been talking with the girl—who in reality was one of our Cyber Crimes investigators—in an online chat room. He even stopped on his way to the meeting and bought some wine coolers to give the girl.

Unfortunately, not all of the people chatting with predators are undercover officers, and not all of the predators are caught in stings.

And not long ago, we arrested a 52-year-old university professor at a bus station in McAllen, Texas, where he was waiting for what he thought was a 13-year-old girl he met online. He even bought the would-be teenager a bus ticket so she could travel from miles away to meet him.

The 80th arrest was particularly notable. It was the arrest of 27-year-old John David Payne, who had been chatting with what he thought was a 13-year-old girl he met on MySpace.

In reality, the graphic sexual conversations he was having were with a Texas Attorney General investigator.

What makes his case particularly frightening—although sadly not unusual—is that, at the time of his arrest, Mr. Payne was out on bail from an arrest that occurred six months prior. In fact, at the time of his most recent arrest, he was already under indictment for online solicitation of a minor. In other words, while he was out on bond awaiting trial for illegal Internet solicitation of a minor, he was back on the Internet, trolling for his next victim. These child predators are dangerous and incorrigible, and children simply cannot be safe with the current landscape of cyberspace chat rooms and social networking sites.

Unfortunately, not all of the people chatting with predators are undercover officers, and not all of the predators are caught in stings. Real children are real victims of real predators. Recently, a 14-year-old girl from Central Texas was raped by a man she chatted with on a social networking site. This is just one of the most recent examples, examples that are repeated around the country with increasing frequency.

Social networking sites and chat rooms have created an environment in which predators target their next victim and plot their next attack.

As we hold this hearing today, millions of teenagers are chatting online, posting personal information on a profile page, talking to other teens on social networking sites, and meeting people in chat rooms. Before we leave here today, countless of those teens will have innocently chatted with someone they didn't know. And, before we leave today, some of those unknown chatters will turn out to be predators who have just located their next target.

Clearly, safeguards are needed at schools and libraries—as well as in our homes—if we are to protect our children against these predators. Such safeguards are the kinds of protections that Americans have come to expect.

Streets, neighborhoods and playgrounds are essential to our daily lives and are part of the American social and economic fabric. Nevertheless, we must police our streets, neighborhoods and playgrounds to ensure their safety. Similarly, the Internet superhighway and social network sites are vital to our modern day economy and they provide an effective platform for the exchange of ideas, information and commerce. They also have developed into virtual neighborhoods where people can simply socialize. But, these modern-day neighborhoods and playgrounds are proving just as susceptible to criminals and predators as their traditional counterparts, if not more so.

As can be expected, the responsibility for protecting children and teens who use the Internet is a shared responsibility. Law enforcement will continue to improve its efforts to track down, arrest and put behind bars anyone who uses the Internet to harm children.

Parents are also a part of the process. Parents must use oversight, education and vigilance to protect their children. Just as parents warn children not to take candy from strangers and to look both ways before crossing the street, parents must warn children about the dangers that exist on the Internet. Parents frequently evaluate whether their teen is sufficiently responsible with a car before allowing him or her to go driving. One reason, of course, is that teens, if driving irresponsibly, could hurt themselves or others. The same is true with the Internet. Parents should ensure that their teenagers are sufficiently informed and responsible in their Internet use. Otherwise, they could get hurt.

But the great weight of the problem must be shouldered by the very creators and hosts of these networking sites and chat rooms that provide the previously non-existent opportunity for child predators. Social networking sites and chat rooms have created an environment in which predators target their next victim and plot their next attack. Predators use these web locations as a starting point for raping a child! The creators and hosts of these networking sites are not the predators who commit the crime, but they create the opportunity for the criminal to carry out his crime.

The creators and hosts of these networking sites are part of the problem, and as part of the solution they must do more than pay lip service to providing a safe environment for children. They must take affirmative, definitive action to ensure the protection of children who use their sites and chat rooms. They can no longer be allowed to turn a blind eye to the predators who lurk on the playground they created.

Admittedly, some of these networking sites and chat rooms have *begun* the process of a constructive dialogue to address the issue. But, if they are honest, they will admit that the predator problem persists, and that there is more they can do to solve the problem.

Should networking sites really mix 13-year-old children with adults who are participating in the networking site for the purpose of looking for a sex partner?

As mentioned earlier, parents have a role to play. But parents are being denied some of their ability to play that role. The networking sites should structure their systems so parents can block access to their sites. Parents across the country prevent their children from going to locations where crime may occur, where sex and drugs may be traded, or where their children could be harmed. Parents should not be denied that opportunity (that responsibility) simply because the location is in a virtual neighborhood, rather than a neighborhood down the street. If parents want their children to have the opportunity to participate on the networking sites, the sites should offer parents filtering software to block their child's access to content and websites that parents deem inappropriate.

Heightened precaution must be taken to protect children under 16 from child predators. Should networking sites really mix 13-year-old children with adults who are participating in the networking site for the purpose of looking for a sex partner? To avoid this problem, the sites should be structured so that only children over a certain age can participate in sites with adults.

Children who use the sites should not be subjected to sexually explicit images and solicitations, or other age-inappropriate material. The sites should be structured so that such material is not accessible without age verification.

While social networking sites are a lot of fun for kids—and have the potential to expose our children to a world of knowledge and bring them literally worldwide friends—many of the sites also subject children to a world of predators, pedophiles and pornographers.

As law enforcement officers, we are asking that social networking sites and chat rooms more effectively police themselves, or be shut down. Our society does not tolerate houses of prostitution. Neither should it tolerate virtual houses that promote predators.

It turns out that the Internet, for all of its benefits and all of its conveniences, is still a pretty dangerous place.

Computer literacy and Internet access are necessities, not luxuries. And without question they have made our lives better. But the anonymity of the Internet has created opportunities for child predators and child pornographers, giving them cover to act on their perversions.

It turns out that the Internet, for all of its benefits and all of its conveniences, is still a pretty dangerous place. It would not be an exaggeration to say that no child is safe from the unwanted advances of chat room predators, men who use the Internet in an attempt to realize their worst fantasies.

Thank you for working to protect children from the nightmare of these predators. We look forward to working with you to win the war against those who threaten the safety of our children, our families, our homes and our communities.

Children Must Be Protected from the Content of Social Networking Web Sites

Elizabeth Racine

Elizabeth Racine is a writer and editor specializing in family and parenting issues.

My name is Elizabeth Racine; I am the mother of three teenagers, and have been a writer, editor, and researcher for the past 21 years, specializing in family and parenting issues. I've appeared in print over 300 times, with articles in *Newsday*, The *Chicago Tribune*, as well as in *Catholic Parent* Magazine, *Today's Catholic Teacher*, and *Momentum* Magazine. I began using the Internet extensively in 1994 for my own research projects as well as for those of my editorial clients. Starting in the year 2000, I taught eighth grade in Catholic schools for four years before returning to full-time freelancing.

As my own children began using the Internet, I have vigorously monitored their online usage, as well as that of my students, with varying degrees of success. Over the past 12 years I have grown increasingly concerned—and now alarmed—about not only what is available for viewing online, especially to our children, but also what our children are posting online as well.

For the past few years I have been giving talks to parents, teachers and clergy about these dangers. My real-life stories and images have shocked audiences, who previously had no idea that this problem existed.

I have served as a resource person for many Catholic reporters, and recently appeared on Fox 29's special report on

Elizabeth Racine, "Submission for the Record of Elizabeth Racine," to U.S. House Energy and Commerce Committee, Subcommittee on Telecommunications and the Internet, July 11, 2006. Reproduced by permission of the author.

Internet abuses by children. Despite the increasing news coverage, there is still a vast amount of Internet activity that goes unreported.

Lately I have been meeting with members of the House of Representatives and the Senate to educate and advise them on these Internet risks to our children, in hope that they will increase legislation and regulation to help protect them.

It is time that Congress steps in to help parents protect their children.

I am grateful for Mike Fitzpatrick's Bill, H.R. 5319, the Deleting Online Predators Act [DOPA] of 2006. While it is not foolproof, it is a starting point, a beacon of hope for those, like me, who have seen the Internet's underbelly and are alarmed over what is happening to our children online.

My goal in writing to you today is to help you better understand this attack on our children. It is my great hope that Congress will assist parents, like me, who are intensely concerned about our children's moral and spiritual development. It is time that Congress steps in to help parents protect their children.

Children See Porn Online

I began using the Internet in 1994, and my interest in the Internet changed abruptly when my then-7-year-old son came to me one day while I was in the garage.

"Mommy . . ." he started, lower lip trembling. "I saw something on the computer . . ." he looked up helplessly when he couldn't find the words. Worried, I asked him to explain what happened.

"I was playing games on Lego.com and then when I was done, I wanted to find more games for boys to play—so I typed in 'boys.com . . .'"

And then he stopped. Seeing his agitation, I ran to the computer to see what would possibly be upsetting him (this was 1997, long before public awareness of rampant pornography). I typed in boys.com and was greeted by an image of a young boy, about 12 or 13, performing oral sex on a man. My 9-year-old daughter then piped in that she decided to check girls.com also, and found an equally appalling photo on that site.

Anyone who has spent any amount of time trolling the pages on MySpace, Xanga, or other social networking sites cannot help but see what a cesspool it is.

Words cannot adequately describe the dark feeling in the pit of my stomach or the primordial mother-bear rage that went through me over this loss of my precious children's innocence. This feeling still wells up in me; my eyes again fill with tears as I write these words. No matter how many times I try, I cannot tell this story without crying, which is embarrassing, even though I have told this story numerous times to parent and teacher groups, trying to get parents to pay attention to what their children are doing online. I endure this willingly in order to protect the children who are being sought so desperately by an industry—indeed, a society—that is *professing* to try to keep them safe, but is more concerned with "free speech" instead of the protection of children. Pornographers and its millions of consumers, are only too happy to snare them into the web of addiction to pornography to keep them coming back for more. Parents and teachers need to be made aware of the need to protect our children, at home and at school and the library, so that their innocent web wanderings don't turn into a nightmare. . . .

Social Networking Sites Are Dangerous

I have researched kids' use of blog sites, now called Social Networking sites, since my own children began using them in

1999. The first ones were the profile sites on Instant Messenger, which could then host "sub-profiles." These sub-profiles allowed children to post all sorts of information about themselves, and I immediately saw the danger in them, since kids could write whatever they wanted on them, and many posted pornography. I tried to alert parents to this danger and met with varying levels of response—some were happy to know so they could correct matters, others said I was out to get their kids and that I was lying about what had been posted! Since I was teaching at the time, this put me in a very awkward position.

With the exponential growth of the latest social networking sites, MySpace being the most popular, I have had my own accounts on all of them, and have been solicited by porn stars—all the big name porn stars try to lure visitors to their own MySpace pages, which feature very suggestive photos, and then, of course, links to the porn site. . . . I received another one today, which invited me to join a MySpace group which is pornographic. I visited the group to see what it was about, and I saw comments posted by lots of kids who were under 18, and who were upset that they had been solicited! So there is hope there, but I can't imagine how many more kids were only too happy to visit the web cam site and participate in the goings-on. . . .

Anyone who has spent any amount of time trolling the pages on MySpace, Xanga, or other social networking sites cannot help but see what a cesspool it is . . . it's no surprise to me that there is an increasing number of child predator solicitations, leading to rape and even murder. The thing I CANNOT believe is that all schools are not required to have a blocking mechanism in place for the students' safety and well-being. Kids are very interested in these sites and very excited to interact with their friends online—and their hormones are raging and most of them cannot pass up the temptation to view porn. They are oblivious to the danger, even when they are educated.

Even from a schoolwork standpoint, as a teacher, I can't imagine trying to teach and having to compete with kids who might be more interested in clicking over to their MySpace account during a class assignment, and having that eat up their academic learning time. And from what I read on MySpace, they do it ALL THE TIME during class. A teacher cannot visually see every computer at every moment, so it is only too easy for kids to do this.

Libraries are another place that blocking software should be mandatory. . . .

Blocking inappropriate content will protect children more than the alternative, which is to allow everything to come to them unfiltered.

DOPA Should Be Implemented

I realize that H.R. 5319 is not foolproof, because blocking technology is not foolproof. Unfortunately, blocking technology can be circumvented, and there are numerous sites being made available to children to allow them to get around their school or home blocking software—these sites are run by adults, which I can hardly believe. But, just because the mechanism isn't foolproof does not mean we shouldn't implement it. Just as seatbelts are not a guarantee that no harm will come to someone who wears one, so blocking technology is to children. Blocking inappropriate content will protect children more than the alternative, which is to allow everything to come to them unfiltered. It is better to do something rather than to do nothing. In the near future, an organization called CP80 (www.cp80.org) will unveil its simple yet ingenious plan to protect our children while it still allows for free speech.

In the meantime, I implore Congress to evaluate seriously these threats to our children and make their protection their foremost concern. H.R. 5319 is a solid beginning, and I ask that it be implemented, along with other bills that seek to protect our children online.

The Deleting Online Predators Act (DOPA) Should Be Passed

Frederick Upton

Frederick Upton is a member of the U.S. House of Representatives from Michigan. He was chairman of the congressional subcommittee that considered the Deleting Online Predators Act (DOPA). The following selection is his argument to the House on the day it passed the bill.

M
r. Speaker, I move to suspend the rules and pass the bill (H.R. 5319) to amend the Communications Act of 1934 to require recipients of universal service support for schools and libraries to protect minors from commercial social networking websites and chat rooms, as amended. [The bill was read aloud by the clerk.] . . .

Mr. Speaker, I rise in strong support of H.R. 5319, the Deleting Online Predators Act [DOPA] of 2006. . . .

The Oversight and Investigation Subcommittee has held multiple hearings exposing the appalling sexual exploitation of children on the Internet. This includes the dark underside of social networking Web sites, which have been stalking grounds for sexual predators who are preying on children all across the Nation; and we have had many and such cases in my home State of Michigan, way too many.

Federal law enforcement officials have described the sexual abuse and exploitation of our Nation's youth as an epidemic propagated by the unlimited access of the Internet. The statistics are alarming. The FBI has seen better than a 2,000 percent increase in its caseload of online sexual predators the last 10 years. And of the estimated 24 million child Internet users, one in five kids has received unwanted sexual solicitations. It

Frederick Upton, "Deleting Online Predators Act of 2006," U.S. House of Representatives, Congressional Record, (H5883-H5889), July 26, 2006.

is estimated that, at any given moment, 50,000 predators are prowling for children online, many of whom are lurking within social networks.

At a minimum, what our hearings have taught us is that both kids and parents need to be better educated about the dangers of social networking Web sites, and parents need to police their children's online use at home to guard against sexual predators.

However, to the extent that children are using the Internet outside the home, particularly at school or at a public library, parents have not been able to monitor their child's online use, and that is the situation that H.R. 5319 is designed to address.

Sexual predators should not have the ability to use our schools or our libraries; and we should take away that avenue, if we can, for their evil deeds.

Earlier this month, the Telecommunications and Internet Subcommittee held a legislative hearing on this bill; and as a result of many constructive suggestions that we heard from our witnesses and Members alike, particularly those representing schools and libraries, the legislation before us today I think reflects much improvement.

Bill Would Require Blocking of Social Networking Sites at Schools and Libraries

At its heart, the bill before us today would require schools which receive e-rate funding [a federal program that gives schools and libraries discounted rates on telecommunications and Internet-service], and I would note that I am a strong supporter of e-rate funding, to enforce a policy of Internet safety for minors that includes monitoring their online activities and the protection measures to protect against access to commercial social networking Web sites or chat rooms, unless used for an educational purpose with adult supervision.

Additionally, this bill would require libraries which receive e-rate funding to enforce a policy of Internet safety that includes the operation of a technology protection measure that protects against access by minors to commercial social networking Web sites or chat rooms unless they have parental authorization and the library informs parents that sexual predators can use those Web sites and chat rooms to prey on kids.

The approach taken by this legislation is not dissimilar to the approach taken by the Children's Internet Protection Act through which Congress requires schools and libraries that receive e-rate funding to impose filtering technology to protect kids from online visual depictions of an inappropriate sexual nature.

Mr. Speaker, I support the e-rate. I continue to do so. I often visit a school, virtually every week. I have seen the tremendous educational value which the Internet has brought to students throughout our district, and I recognize the importance of the e-rate in making that a reality.

However, as with all technologies, the Internet is a double-edged sword, and Congress does have the responsibility to ensure that, to the extent that a Federal program is involved, like the e-rate, it is doing all that it can to ensure that children are protected from online dangers. This bill represents another step in making sure that online experiences at school and the library are safe, . . . [debate by other members of Congress followed.]

The Bill Has Many Supporters

This is not the end-all. We know that. But we know that sexual predators should not have the ability to use our schools or our libraries; and we should take away that avenue, if we can, for their evil deeds. And that is precisely what this legislation is intended to do.

And I would note that even though this was introduced some 2 months or so ago, we have nearly 40 cosponsors of the legislation. Melissa Bean has been a great leader from the Democratic side of the aisle, as well as the Republicans that have been mentioned earlier during the debate and that have participated. And I know that in the oversight and investigation hearings that we have had, not only as well as in New Jersey but the legislative hearing that we had with many witnesses, including the Attorney General from Texas, who did a marvelous job of explaining what was going on in Texas, they all strongly endorsed the intent and the legislation as it was introduced. . . .

With school starting for many as early as next month, August, knowing that the Congress, at least the House, is likely to adjourn this week, the Senate still has another week, I would like to think that with a strong vote this afternoon the Senate may take up this legislation perhaps next week, perhaps, and we actually may get the bill to the President's desk so that it will be in place for kids as they start school.

So that is one of the reasons, I think, why this legislation was, indeed, rushed to the floor. But, again, I know that we took in many good comments by those at the legislative hearing that we had, and I think that proof will be in the pudding. . . .

Mr. Speaker, I would say, too, I believe it was over the Fourth of July week break that *Dateline* had the big expose; and I know our office was flooded with calls and letters, as I was home in Michigan. There were a good number of parents and others that expressed their concern about some of these different online services that were there; and if we can close the loophole on schools and libraries, I think that it is a good thing.

I think that, because of that, I would hope that most Members, when we vote on this later this afternoon, in all likelihood would vote "yes" on the bill. Again, it is bipartisan, and that is why it is here before us this afternoon.

Blocking Social Networking Web Sites Would Not Make the Internet Safer

Michael Resnick

Michael Resnick is associate executive director of the National School Boards Association. The following selection is a letter he sent to members of the U.S. Senate.

On behalf of the National School Boards Association (NSBA), which represents the nation's 95,000 local school board members, I would like to alert you to our opposition to H.R. 5319, the Deleting Online Predators Act (DOPA), which may be placed on the Senate calendar for consideration next week. This legislation would require schools and libraries to block access to social networking sites as a condition for receiving E-rate funds.

As you know, the issue of social networking has gained attention in recent months due to several reported cases of sexual predators pursuing children online through interactive web sites. NSBA deplores such actions and believes that these violators should be prosecuted and punished to the full extent of the law. Recent publicity of these cases, however, should not rush us into enacting bad policy that would be detrimental to the future of online learning.

Although this legislation may look attractive inside the beltway, in the real world, it has no credible support. In fact, school board members, school administrators, libraries, the U.S. Chamber of Commerce, education technology groups, and parent groups such as the PTA all oppose this legislation.

In its current form, the Deleting Online Predators Act will not make the Internet a safer place for America's schoolchildren. Even the Center for Missing and Exploited Children indicates that "more restrictions will cause teens to go somewhere else that has fewer restrictions, with the unintended consequence of increasing their chances of being victimized."[1] Instead, the bill unfairly targets schools and libraries by placing unnecessary and costly requirements on them. Furthermore, the legislation does not address the real issue of educating children about the dangers of the Internet and how to use it responsibly and wisely. Information and communication technologies are part of a meaningful and relevant education and are essential tools in preparing students with the skills that they will need to be competitive in the global marketplace.

In addition, it should be noted, that the Children's Internet Protection Act which Congress adopted in 2000 already requires school boards, schools, LEAs or other school administration authority to certify that they are enforcing a policy of Internet safety for minors and to use technology to protect against obscene or harmful material. As a result, school districts have the power to block access to social networking sites and chat rooms, and a number of them have chosen to do so. A federal mandate requiring all E-rate applicants [schools that receive discounted rates on Internet access] to block social networking sites, would impose additional costs on schools, and usurp local districts authority to determine what content should flow into schools. These consequences could deter schools from participating in the Internet access program altogether.

The proposed ban on social networking sites could eliminate some very innovate practices in schools such as teacher use of blogs. For example, one educator in Liberty, MO uses a blog and podcast for American history lessons and has had downloads from around the world. Another teacher in Orange

---

1. Testimony of Michelle Collins, National Center for Missing and Exploited Children, U.S. House of Representatives, Energy and Commerce Committee, Subcommittee on Telecommunications and the Internet, July 11, 2006.

City schools in Ohio uses a blog for 4th graders to become "book critics" by posting their book reports online.

While DOPA allows for the disabling of blockage software by adults or minors with adult supervision for *educational* purposes, this process is unrealistic and would be cumbersome, costly, and difficult to enforce. Unlike current restrictions against obscene materials that can be objectively identified, this legislation would require schools to subjectively predict which sites may be misused. Identifying and evaluating such sites are not compatible with the technical capabilities of most filtering vendors and are likely to lead to blocking of legitimate instructional sites. Furthermore, there are no resources contained in the bill to assist schools in purchasing such software if it did exist.

Perhaps most importantly, however, the blockage of social networking sites while children are at school does not adequately prepare them to deal with the potential dangers of online predators outside of school and how to develop responsible Internet usage skills. Software filters and other blockage devises provide a false sense of security for America's children. Whether students are able to access these sites at school or not, they still need to know the right way to navigate and conduct themselves in a Web 2.0 environment, where they are now the creators of web-based content. As the National Center for Missing and Exploited Children contends, "We need to help protect children through education, open dialogue and elevated awareness."[2]

According to recent data, 87% of students 12–17 are online and 65 million young people have accessed social networking sites. Given this reality, we cannot afford to "duck" our responsibility to educate our children by hiding behind filters to block these sites. Education is the key to preparing our students to interact safely on the Internet. Together, the education community, parents, and others can help to instill

2. Ibid.

responsible decision-making and empower students to recognize the lures of online predators.

On behalf of school board members across the country, NSBA encourages you to oppose this bill if it comes to the Senate floor. At the very least, this legislation should have a chance to be considered by the Senate Commerce, Science and Transportation Committee, which has jurisdiction over this matter, for more thorough examination of the policy implications of the bill.

Thank you for your consideration. We look forward to working with you to ensure that America's schoolchildren receive a safe and valuable educational experience. Should you have any questions about this issue, please contact Chrisanne Gayl, Director, Federal Programs, at (703) 838-6763 or cgayl@nsba.org.

Blocking Social Networking Web Sites Would Hurt Poor Children

Sonia Arrison

Sonia Arrison, a TechNewsWorld *columnist, is director of Technology Studies at the Pacific Research Institute in California.*

Social networking Web sites like MySpace, Friendster and Facebook are becoming increasingly popular with the nation's youth, prompting attempts to control the medium. However, though protecting children is the goal, the outcome is too often the opposite.

Recently proposed legislation by Congressman Michael Fitzpatrick, a Pennsylvania Republican, threatens to effectively stop minors from accessing social networking sites in schools or libraries. His Deleting Online Predators Act (DOPA) would prohibit schools and libraries from allowing access to a commercial social-networking Web site or chat room through which minors might be subject to sexual material or advances. For many kids, that amounts to a ban on using the sites from anywhere outside the home.

Spreading Safe Practices

It's true that the crowds of young, impressionable people on social networking sites have attracted child predators like bees to honey. Indeed, just last month Americans were shocked to discover that the deputy press secretary at the Department of Homeland Security, Brian Doyle, was arrested during an Internet sex sting. Doyle was a high-profile individual—and there are others like him.

Child predators are a real and serious problem, but the answer isn't to turn off access for legitimate users. A better approach is first to refocus law enforcement so that the Internet is a normal place of patrol, and, second, to engage parents and businesses.

Those supporting DOPA might want to consider that poor children rely more heavily on school and library Internet access than do their wealthier counterparts.

The best way to stop child predators is to teach kids how to protect themselves online. Parents and educators have a key role to play and numerous non-profits exist, such as cyberan gels.org and staysafe.org, to help spread safe practices. Children need to know that dangers exist, and more importantly, they should practice avoiding them.

It is incredibly important for kids to be safe online, but when laws like DOPA threaten to hinder communications in order to stop a few bad apples, everyone loses. Recall the fallout after the implementation of the Children's Online Privacy Protection Act (COPPA) in 2000. That law requires Web sites to obtain "verifiable parental consent" before collecting personal information from children under 13 years old.

The purpose was to protect kids, but instead the law created negative unintended consequences by cutting off good communications. For instance, the popular television show "Thomas the Tank Engine" announced it would stop regular e-mail bulletin service because of the law and NBCi decided to close all e-mail accounts registered to kids under 13. There are many other cases of opportunities closed off, especially for poorer kids, because of the law.

Empower the Individual

Those supporting DOPA might want to consider that poor children rely more heavily on school and library Internet access than do their wealthier counterparts. Social networking sites have become so popular that they are a significant tool for socializing and learning from peers. Take that away from the underserved communities, and that will be one more setback that will cause all sorts of problems, including crime, in the future.

Children need to be safe online, but the way to ensure their safety is not by cutting off their access to virtual playgrounds. In a free and open society, the best way to guard against threats is to empower individuals—especially children and parents. If kids do not gain experience in protecting themselves early on, they won't be able to do it later. Lawmakers who truly want to protect kids should reject any proposals to ban their access to technology.

Blocking Social Networking Web Sites Would Deprive Children of Educational Opportunities

Beth Yoke

Beth Yoke is executive director of the Young Adult Library Services Association (YALSA).

Chairman Upton and Members of the Subcommittee, thank you for inviting me today to testify on behalf of the American Library Association (ALA). I sincerely appreciate the opportunity to comment on H.R. 5319, the Deleting Online Predators Act (DOPA). ALA has three primary concerns about DOPA: 1) that the broad scope of this legislation will limit access to essential Interactive Web applications; 2) that the legislation would widen the digital divide by limiting access for people who use library and school computers as their primary conduits to the Internet; and 3) that education and parental involvement are and have always been the best tools to keep kids safe online and to ensure that they can make the right decisions.

I am the Executive Director of ALA's Young Adult Library Services Association (YALSA). The American Library Association is the oldest and largest library association in the world with some 65,000 members, primarily school, public, academic, and some special librarians, but also trustees, publishers, and friends of libraries. The Association's mission is to provide leadership or the development, promotion, and improvement of library and information services and the profession of librarianship in order to enhance learning and ensure access to information for all.

Beth Yoke, testimony before the U.S. House Energy and Commerce Committee, Subcommittee on Telecommunications and the Internet, July 11, 2006. Reproduced by permission of the author.

Before taking this position with ALA, I was a young adult services librarian and an instructor of future school library media specialists and youth librarians at West Virginia University. I can say with authority that no one is more concerned with the safety of children online than librarians—especially youth librarians.

DOPA uses the term "social networking sites" in an overly-broad way to describe virtually all Interactive Web applications in which users converse or interact with each other.

Youth librarians believe, and more importantly know from experience, that education about safe Internet practices—for both youth *and* parents—is the best way to protect young people. We believe that the overly broad technological controls that would be required under DOPA are often ineffective given the fast-moving nature of modern technology. Further, such technological controls often inadvertently obstruct access to beneficial sites. In essence, we believe that this legislation will lead to the blocking of essential and beneficial Interactive Web applications and will further widen the digital divide.

The Terminology Used in DOPA Is Flawed

It is very difficult to define many of the terms used in the debate over Internet usage. DOPA uses the term "social networking sites" in an overly-broad way to describe virtually all Interactive Web applications in which users converse or interact with each other. As it is currently written, the definition (even with the educational exemption) would include: educational tools used to provide distance education, community forums that allow children to discuss issues of importance, online email programs through which family members can communicate with each other and with teachers and librarians at their local schools and libraries and even find one another in cases of emergency. There is enormous value to be found in these interactive online environments. Blocking access to them denies young people the opportunity to benefit from all the Internet has to offer while not necessarily ensuring kids' safety online. As written, HR 5319 is simply too broad and would block access to many valuable applications. We urge you to consider changes in the bill language.

There are many examples of online education applications that would be blocked under DOPA. One example reported in *Education Week* indicated that more than 10 million students were part of an online field trip to the Carlsbad Caverns National Park in Carlsbad, N.M.—without leaving their classrooms.

There are countless positive uses for networking applications that are not necessarily related to formal education.

The April 25 field trip, coordinated by Ball State University [BSU] in Muncie, Ind., was to consist of two live virtual tours of the cave featuring scientists, park guides, and First Lady Laura Bush. Students in grades 3–8 were invited to call in or e-mail questions to be answered on the air, or to participate in an online discussion during the 90-minute broadcasts. The organizers billed the event as the largest simultaneous visit ever to a national park. It was also described as the largest "electronic field trip" ever broadcast by BSU, which has organized more than 50 such trips since 1996.

Use of the site WebCT.com, an online education application, has helped thousands of people get their degrees through distance learning. WebCT is a site that allows users to enroll in and participate in classes online. It can also act as a forum for class discussions. Users create profiles and become students in a virtual classroom through online bulletin boards, real-time chat, student blogs, and more. In the rural areas of states like Texas, Wyoming, the Upper Peninsula of Michigan and West Virginia—where I worked, distance from major hubs once created an enormous barrier to learning.

Today, applications, like those available through WebCT, make it possible for young adults to complete degrees online at schools that are hundreds of miles away as well as access other non-credit courses for personal interests. It is our analysis that DOPA would make this type of distance learning impossible, since the bill requires that interactive applications, like courses available through WebCT, can only be used in libraries or schools if there is adult supervision. But what does "adult supervision" mean? Or how is "education" defined in the pertinent exemption? Is it formal for-credit only courses? And, why would we create barriers for young people who want to take advantage of online educational opportunities?

There are countless positive uses for networking applications that are not necessarily related to formal education. Networking applications include support groups for teenagers with physical or emotional disabilities, forums for the exchange of ideas, and even tools to help kids become acclimated to new surroundings. For example, when teenagers leave for college they often use networking sites to find other students with similar interests.

Finally, with specific regard to "social networking sites," young adults all over the country have begun to use these sites as a primary means of communication, whether with their peers or with young adult authors, musicians, artists, and with libraries. Some libraries are taking advantage of this by using some of these sites to stay in touch with their communities. For example, Sean Rapacki from the Wadsworth Public Library in Wadsworth, Ohio informed us that his staff has created a MySpace profile page FOR the library, allowing library staff to communicate with young adult patrons much more effectively. [By 2007 hundreds of public libraries had MySpace profiles.]

DOPA Ignores the Value of Interactive Web Applications

Today's interactive online environment is an essential and growing part of economic, cultural, civic, and social life. New Internet-based applications for collaboration, business and learning are becoming increasingly important, and learning to use the online environment effectively and safely is now an essential component of education.

Why would you limit access to interactive Web applications in the one place where students can learn to use them safely?

The popularity of interactive online environments is extending to the corporate world as a number of businesses—including corporate leaders like Ernst & Young and IBM—are increasingly moving to the use of interactive Web applications as a primary mode of business communication. These companies recognize that technology can be an essential way to achieve a home/work balance and maximize efficiency. As the corporate, professional, and civic use of interactive Web applications grows, there is an increasing need for today's young people to be prepared to thrive in a work atmosphere where meetings take place online, where e-business is a driving force behind the world economy, and where online networks are essential communication tools. This is exactly what kids must be ready for: a professional environment where only the web-savvy thrive.

The Internet is changing how we live, learn, work, and interact with one another. If today's young people are to succeed in the workplace of the future, they must learn information literacy skills for the technologies of today and tomorrow. Libraries are far and away the best places to learn these skills, and social networking sites, which *introduce* kids to the world of online interaction, are key to successful development in that field.

Education, Not Laws Blocking Access, Is the Key to Safe Use of the Internet

Libraries and schools are the locations where students develop the information literacy skills that are vital to success in today's world. Information literacy includes "the abilities to recognize when information is needed and to locate, evaluate, effectively use, and communicate information in its various formats." These are skills that public librarians and school library media specialists are in a unique position to foster in young people. In today's world, information literacy skills go far beyond computer instruction and web searching. In a fully developed information literacy program, students learn how to find, evaluate, and use online information and also learn how to use critical thinking skills to make good choices online.

This type of information literacy instruction is going on in schools and libraries all across the country. Legislation like DOPA sends exactly the wrong message at the wrong time— instead of allowing librarians and teachers to instruct students about how to use all kinds of applications safely and effectively, it creates barriers to information literacy instruction. This flies in the face of hundreds of years of educational theory—why would you limit access to interactive Web applications in the one place where students can learn to use them safely?

Fortunately, thousands of public and school libraries across the country, along with websites like Ms. Aftab's WiredKids .org, are doing an outstanding job in helping parents teach children how to use the Internet safely and responsibly. For example, Baltimore County Public Schools and the Baltimore County Public Library co-sponsored a *Family Guide to Child*

Safety on the Internet, a valuable resource for helping parents teach their kids the do's and don'ts of web surfing.

Local Decision Making—Not Federal Law—Is the Way to Solve the Problems Addressed by DOPA

As advocates for effective use of information, librarians and teachers are fully committed to helping young people have safe online experiences; furthermore, we teach young people the information literacy and critical thinking skills they need to use the Internet safely and effectively. We reach the goal of educating kids to safely use online tools with information literacy education programs as described above, and through locally-developed online safety policies, which in many cases include the use of technological barriers like filters. In other words, many of the problems that DOPA seeks to remedy are already addressed at the local level.

Research shows that use of the Internet, including interactive sites, leads to improved reading and other academic successes.

About 80% of all public library funding is local, library programs are developed to be responsive to local requests, and the policies governing libraries are developed with local trustees and community members; comparable policy decisions are made by school boards. Federal legislation like DOPA robs libraries and communities of local decision-making and control. Decisions about what is best for an individual community should be made by the community itself.

DOPA Would Restrict Access to Technology in the Communities that Need Public Access Most

According to recent statistics from the U.S. Census Bureau the digital divide is large and does not appear to be shrinking.

Currently, roughly one out of seven African Americans and only one out of eight Hispanics subscribe to broadband service at home. Meanwhile, 26.1% of whites and a full one third of Asians have broadband at home. Further, according to a Gates Foundation report on the role of libraries in Internet access, African Americans and Hispanics "rely exclusively on the library computer for Internet access to a greater degree than their white and Asian counterparts," with approximately a fifth of African American users and nearly 16% of Hispanic users exclusively relying on library-based computers. Also, nearly a third of those in the lowest income bracket who use library computers rely exclusively on them. These data indicate that public access computing in libraries is playing an important role in closing the digital divide.

Since DOPA, as presently drafted, would require libraries and schools receiving E-rate discounts through the Universal Service Program to block computer users from accessing interactive Web applications of all kinds, opportunities for those who do not have Internet access at home would be further limited.

There is a great deal of research being conducted on young people and their use of the Internet and interactive applications. There is much to learn from this research about educating young people and helping them to safely use the Internet. For instance, research shows that use of the Internet, including interactive sites, leads to improved reading and other academic successes. Sadly, research reported by Bruce Bower indicates that "children most likely to benefit from home Internet access are the very children least likely to have [it]."

The ALA would like to affirm the importance of online interaction and collaboration and the development of essential information literacy skills. We at ALA stand ready to work with you, to assure that our children are protected, educated, informed and made as safe as possible.

Laws Should Target Sex Offenders, Not Web Sites They Visit

Adam Thierer

Adam Thierer is a senior fellow with the Progress & Freedom Foundation (PFF) and the director of PFF's Center for Digital Media Freedom. He is the author or editor of five books on technology and media policy.

I want to begin by saying a few words about the beneficial side of social networking. What troubles me most about the current debate over social networking websites is that almost no one is talking about the beneficial side of social networking, and indeed the Internet as a whole, for children.

For the vast majority of civilization, humans have lived in a state of extreme information poverty. Today, by contrast, we are blessed to live in amazing times. An entire planet of ubiquitous, instantly accessible media and information is now at our fingertips. We are able to share culture and learn about other cultures in ways that were unthinkable even just 10 years ago. And social networking sites are one of the most exciting components of this new interactive, borderless world. Indeed, they are sort of the equivalent of "digital town squares" where citizens—including children—can gather and communicate with each other.

True, some bad guys might gather there too, but that is the case with every type of town square when you think about it. Thus, instead of only focusing on the negative, lawmakers need to acknowledge all of the wonderful things that happen on these sites too.

Adam Thierer, "Remarks at Conference on Social Networking & Child Protection," *Technology Liberation Front*, June 23, 2006. www.technologyliberation.com. Reproduced by permission of the author.

Enforcement Challenges, Part I: Internet Realities

But let me turn to some of the enforcement challenges law-makers will face, and are likely ignoring, when they propose regulating social networking websites in an effort to protect children from online predators or other threats.

First, I hope policymakers understand that it will be impossible to censor or shut down social networking sites entirely. When you think about it, the entire Internet is really just one big social networking site. Any personal website can be a social networking site. Blogs are certainly social networking sites, too. Hobbyist sites, list-serves, bulletin boards and chat rooms are the oldest forms of social networking websites. But even e-mail and instant messaging are forms of social networking. And massive, multi-player online video games are quickly becoming one of the most popular forms of social networking.

Kids have been abducted in or near shopping malls in the past. But you don't hear anybody proposing a ban on shopping centers or convenience stores to solve this problem.

Thus, social networking is everywhere; it is the Internet. What this means for purposes of our discussion here today is that there will always be another site for the bad guys to flock to. Why, therefore, are we singling out social networking sites like MySpace.com and others? If we're going to regulate them in the name of protecting children, then we'll need to regulate a lot more Internet sites and online activity to adequately do the job. Rep. Michael Fitzpatrick's bill [the Deleting Online Predators Act], for example, would ban social networking sites in schools and libraries. But what will that really accomplish if every other sort of social networking site and service is left unregulated?

Moreover, singling out online social networking sites and services strikes me as peculiar since we don't also single out offline social networking sites for special regulations just because children might be at risk there. A lot of kids hang out at shopping malls, for example. And, unfortunately, we know that kids have been abducted in or near shopping malls in the past. But you don't hear anybody proposing a ban on shopping centers or convenience stores to solve this problem. Such a proposal would strike most people as absurd; even outrageous. But, in essence, that's what lawmakers are doing when they single out social networking websites for unique treatment or regulation.

Enforcement Challenges, Part II: Age Verification

I want to also address the argument that social networking websites should just start age-verifying all their users to ensure better security or online safety. In reality, there are no perfect solutions to the age-verification problem because, at root, it is a human verification problem. In order to firmly establish someone's identity many different pieces of information about that person are necessary.

And with teenagers (especially under the age of 16), this is extraordinarily difficult. They are not voters. None of them have home mortgages or car loans. Many of them don't have a driver's license. Most of them are not in the military. Those are methods we use to identify adults that generally won't work for teens. Moreover, the few records we have on kids are well-guarded. For example, their school and health records are not publicly available. Same goes for their social security numbers.

In other words, unless you want to propose national ID cards for kids, we do not have an effective way to age-verify kids online. And credit cards cannot serve as a rough proxy of age ID in this case like they do in other online contexts because (a) most kids don't have credit cards; (b) even if kids had credit cards, they wouldn't need them to use social networking sites or services since they're not buying anything most of the time; and (c) there's always the chance that if they wanted to get on social networking sites bad enough kids would find other ways to get credit card numbers and go online under false pretenses.

How much time, energy and resources are we putting into actually trying to track down the real bad guys as opposed to regulating websites?

Thus, we have to accept that kids are going to be online and that we're not always going to be able to perfectly identify them when they are online. Consequently, we're going to have to redouble our efforts to teach our children basic rules of safety both online and offline and continually remind them to jealously guard their personal information and pictures, not to meet up with strangers in public places, and to always talk to Mom and Dad about their concerns or questions.

Enforcement Challenges, Part III: Are We Addressing the Real Problem?

I want to conclude by asking a provocative question about protecting children from online predators: How much time, energy and resources are we putting into actually trying to track down the real bad guys as opposed to regulating websites? Regulating social networking strikes me as a massive misallocation of resources and attention.

Indeed, let me pose this question in this fashion: Is this a case of market failure or government failure? Consider this fact: A 2003 Department of Justice study reported that the average sentence for child molesters was approximately seven years and, on average, they were released after serving just three of those seven years.

Now let me just ask—no, let me scream it at the top of my lungs—WHY ARE THESE CREEPS WALKING THE STREETS? Why are we only putting people who viciously hurt innocent children behind bars for just seven years and then letting them out after just three? This is insane!

And this is what gets me so incensed about the current debate over social networking websites: *Policymakers are fond of pointing fingers at everyone else and scolding them for not doing enough to protect children from predators, all the while conveniently ignoring their own policies that allow those predators to be on the streets and behind keyboards in the first place!*

What's even more troubling about this is that, after letting the child abusers out of jail, we then plow a lot of money and law enforcement resources into "community supervision" and "sex offender registries" to give us a better idea of where all the child molesters live in our neighborhoods. Well, I don't want to say that I speak for all parents out there but I bet I speak for a lot of them when I say the heck with all these "community supervision" programs and "sex offender registries" because I don't want these scumbags on the streets and anywhere near my son or daughter! I find it very troubling that I have to go onto the sex offender registries and find all the convicted child molesters living in my neighborhood and then figure out how to keep my two kids away from them. Call me old fashioned but I wouldn't mind the old "lock-'em-up-and-throw-away-the-key" approach with these vermin who prey on our children. At a minimum, we ought to be considering sentences that are a heck of a lot longer that just seven years in which they let get out after doing just three years of hard time.

So, again, is it market failure when we hear that a bad guy is lurking online at a social networking site, or is it government failure? I would suggest it is government failure in the extreme because if those bad guys committed crimes before then they should probably still be sitting in a jail cell instead of in front of a keyboard trying to lure our children in.

And even if they haven't committed a crime against children before, the government should be using its resources to find those who attempt to lure our children and see if they can catch them in a sting operation before they harm our children. Social networking websites will be happy to assist in that effort when a potentially threatening individual is identified or suspected. MySpace.com and other sites already have firm policies in place of working with law enforcement to identify threats and take action against them. After that, it's up to law enforcement to do the right thing and put these predators away for a long, long time.

I would hope you would agree that this represents a superior approach to this problem compared to proposals to censor or regulate social networking websites because I really don't believe that approach would do much to protect our children from online threats.

Online Social Networking Should Not Be Feared

Vicki A. Davis

Vicki A. Davis is a teacher, conference presenter and writer who maintains several blogs, including the Cool Cat Teacher Blog.

Consider the lobster. When it is in warm water, it frequently molts its shell. A new, very soft shell grows underneath its new one. When it is time, the lobster begins to literally shrink as it expels seawater and the old shell begins to split.

After going through this gruelling process, a lobster emerges with a new soft shell. With a soft shell, it must literally hide in its borough for one to two weeks or it will be easy prey for fish and other predators.

Growth and molting are an important part of the growth process of the lobster with even the eyes of the lobster shedding their covering. Without molting, the old shell would become the lobster's coffin.

Growth and Change Are Part of Life

Change is the only constant in our world. In America, in the warm waters of democracy and prosperity, things change rapidly.

It requires us to shed our old modes of thinking and to adopt new methodologies and paradigms in order to understand the new world. We must understand, act, and adapt to the changes. We must comprehend the changes so that we, as educators and parents, can teach our children morality and ethics in light of the new societal shift. To ignore and refuse to adapt to the societal shift results in lost opportunities and wasted resources.

Vicki A. Davis, "Societal Shift and DOPA," *Cool Cat Teacher Blog*, July 30, 2006. http://coolcatteacher.blogspot.com. Reproduced by permission of the author.

Congressmen and Senators are extremely aware of the "New" Internet and honestly I don't think they like it. When Trent Lott became the first Senate Majority Leader in history to resign under pressure, they became acutely aware of blogs.

As an educator, I want to harness social networking to create online islands where teens share educational interests.

Anything that takes the common person, like me, and gives them a voice, unnerves those who are in power and do not want to change and adapt. It also unnerves those who have chosen obsolescence as a mindset.

Social Networking and Societal Shifts

In my parents' teenage years, they congregated at hamburger stands and drive-in theaters. I remember them telling me that the parents used to complain about all of the problems there. My generation gathered at the mall or at local fast food restaurants after sporting events. Many liked to "drive around" in local spots. Again, parents complained.

Today's kid doesn't go to the mall or "drive around" as much as they congregate online. That is where they converse. When we "drove around", the parents responded by having more policemen put on duty at the places we congregated. They were a safety net. Kids who want to "drive around" are going to "drive around" and the police kept away the problem folks.

We should respond by creating visible presences of "online safety police" and reporting mechanisms for predatory behavior. We should respond by educating children and parents both at home and from school about the dangers. We should teach parents how to look up and monitor their children's myspace accounts.

Do Not Be Afraid of Social Networking

Social networking is not to be feared, a recent Pew study found:

> Our evidence calls into question fears that social relationships—and community—are fading away in America. Instead of disappearing, people's communities are transforming:

> The traditional human orientation to neighborhood- and village-based groups is moving towards communities that are oriented around geographically dispersed social networks.

> People communicate and maneuver in these networks rather than being bound up in one solidary community. Yet people's networks continue to have substantial numbers of relatives and neighbors—the traditional bases of community—as well as friends and workmates.

Websites and the Internet are a conduit and must not be confused with the creeps harming our children!

Transformation is a process. Like the lobster, we can transform. We can molt and shed our old shell and create new methodologies of protection that work well with the online world.

As an educator, I want to harness social networking to create online islands where teens share educational interests. History buffs, science prodigies, math talents, literature lovers! Students often rise to the level of those that surround them. If we want students to rise to new heights, we will create conduits for educational exchange.

I fear that DOPA [the Deleting Online Predators Act] will derail these efforts for at least several years if not longer, depending on how the law is enacted. And that is just it. No one can count the number of sites on the Internet, how is a commission going to sort through every one of them?

Regulate the People, Don't Cut Off the Conduit

Websites and the Internet are a conduit and must not be confused with the creeps harming our children!

Perhaps a telephone repairman should look at a bundle of fiber optic cabling and listen in on every phone call and cut the cables with offensive conversations? We wouldn't dream of it! It is not the cabling that is the enemy, it is the people on each end of the line.

This is not about shedding morals but rather taking morals into a new online world through education. Our country must change, shed old ways of doing things lest our old ways become our coffin in a new society bursting to be released from its shell.

Why Isn't the IT Industry Screaming?

I think more IT folks are not screaming because I think this bill spells big dollar signs for them:

- More money for phone companies who will have to provide capacity to the centralized "content filtering database" that will ultimately be created at the Federal Trade Commission

- More money for in house servers to set up internal blogs and wikis

- Money to replace the online free services that schools use with internal servers, equipment, and manpower

- More money for consultants

- More money to hire more bureaucrats.

And as a result we have students and parents who are no more educated about living in an online world than when this all started. . . .

I believe that we are bordering on becoming a non-adaptive society as Congress goes ahead with the DOPA act.

What We Need to Do to Protect Children

- Yes, all schools must have filtration that protects children from obscene, pornographic, and dangerous material. It should be controlled locally so that it can adapt quickly to local teen issues and curriculum needs. (This could actually be verified remotely by audit by the FTC if it was needed. The technology exists today to do this!)

- Yes, all students must be supervised with school activities. Teachers should be allowed to supervise using RSS feeds from student created materials with the realization that when one deals with teenagers and technology that a "zero defects" result is unrealistic. Issues should be dealt with using an effective discipline ladder tailored to electronic issues. Discipline should include suspensions of student accounts, and alternative offline assignments.

- Parent education programs should be created to run parallel to student programs. (I offered a course in 1997–1999 at our school that parents and students took together and ended with a parent/student contract for online safety. I am updating it and bringing it back this fall.)

- Students should be educated about information literacy, online safety, and online privacy and should be monitored by parents and teachers to make sure they are following such procedures.

- A mechanism for reporting online predatory behavior must be created with law enforcement dedicated to policing such behavior.

- Yes, all schools should fully disclose the Internet activities that children are doing at school. (I post them on our school website and invite parent comments.)

- Yes, we must teach children responsible, ethical use of new Internet tools in a way that will best allow students to succeed when moving to the real world.

- I would propose that all ISP's be required to provide free content filtration for parents as well as a summary printout to parents of websites that their children go to. This could be done and provided for school and for home. I do not think a child should use the Internet without some sort of filtration in place. Now, that is something that would help the problem!

I also think advocates of DOPA should listen to educators who have their finger on the pulse of what is happening with children and understand that DOPA falls far short of providing a safety net for kids.

In fact, ignorance is far more dangerous than supervised education. We must learn to adapt to the fundamental societal shift that has occurred as we begin to live online in ways that will protect our children today and their future success tomorrow.

For Now, We Refuse to Change

Fighting change for the lobster means death. Fighting change for us means being out of touch with the things we can do that can really help our kids and keep them safe!

As for now, I think the shell is tightening.

This quote has been floating around the net:

Filter a website and protect a child for a day.

Teach them online safety in a near-real world environment and protect them for a lifetime.

CHAPTER 3

Who Is Using Online Social Networking?

Chapter Preface

Much that is written about social networking Web sites gives the impression that they are used mainly by teenagers, and that all teenagers are interested primarily in sex, dating, or trivial chatting with their friends. But this is far from the truth. In the first place, people of all ages use social networking sites. Even at MySpace, which is best known to the public for its popularity among teens, the proportion of older users is large and growing. And in the second place, both teens and adults do a great deal more on MySpace than simply socialize.

According to an August 2006, ZDNet survey, the percentage of MySpace visitors under 18 had decreased from 24.7% to 11.9% in the past year and 40% of visitors were aged 35 to 54. However, it is impossible to get accurate statistics on the ages of MySpace users because so many people give false birthdates—not just young teens claiming to be older, but adult women who choose not to reveal their true ages and organizations to which age is not applicable. Thus there are a great many accounts showing ages over 80, most of which are arbitrary, although some senior citizens do use the site.

As of July 2007, MySpace had over 190 million registered accounts, and the number grows by millions every month. But this does not mean there are 190 million individuals logging on. Some accounts are inactive, while many people have more than one account. In addition, organizations—both commercial and nonprofit—have their own profile pages. For example, many public libraries, especially their teen departments, are on MySpace. Hundreds of advocacy organizations attract supporters and organize action through their MySpace pages; environmental causes are especially prevalent, but plenty of others are represented from across the political spectrum. And of course, there are pages not only for bands, but for popular TV shows and movies, as well as authors, publishers, and an increasing number of retailers.

Why would an individual have more than one MySpace account? Some people have both a personal account and one related to their professional work. A great many have accounts for their pets—for instance, a lot of cats have MySpace profiles, to the delight of cat lovers who enjoy sharing pictures of them. Authors sometimes create pages not only for specific books but for characters in their books. There are also unofficial profiles for such entities as sports teams, cities, and planets.

But the most notable use of "extra" profile pages by individuals is the proliferation of fan pages and tribute pages. These profiles carry the name and picture of a celebrity or other famous person and contain information about that person, as well as photos. While many such pages are focused on entertainment stars, there are also many that honor people in other fields, including writers, scientists, astronauts, politicians, and historical figures. Some observers feel that people should not post profiles under "fake" names, but most fan and tribute pages do not contain any element of pretense—they are often written in the third person and represent a creative effort to describe someone's contribution to the world. These pages collect hundreds of "friends" among users who want to express their admiration for the person honored.

The lesser-known uses of MySpace balance the negative ones that have received so much publicity. There is, to be sure, pornography there, but it is by no means dominant among the great variety of profiles that exist. The authors in this chapter describe only a few of the things, both bad and good, that online social networking is used for.

Sexual Predators Find Victims Through Online Social Networking

CBS News

This report was aired on the CBS evening news.

It all started on the social networking Web site MySpace.com. A 14-year-old girl began receiving graphic messages from a much older man, asking whether she was "OK with me being 38?"

It wasn't the first time the alleged predator, Robert Wise, trolled the Internet looking for sex, according to Sgt. Dan Krieger of the League City, Texas, police department. But this time, reports CBS News Correspondent Sandra Hughes, the authorities stepped in.

"We assumed her online identity and started chatting with this guy," Krieger explains. "During that point, he made it very clear he wanted to meet her for sex. We were able to find another 14-year-old female that he's actually had sex with."

Wise is now in custody, charged with multiple counts of sexual assault.

But the incident is just one of many cases nationwide—and some of them have ended tragically.

Possible Murder Connections to MySpace

In New Jersey, Majalie Cajuste is grieving the murder of her daughter Judy. The 14-year-old reportedly told friends she met a man in his 20s through MySpace.com.

Across the country, in Northern California, friends are mourning 15-year-old Kayla Reed. She was active on MySpace until the day she disappeared.

CBS News, "MySpace: Your Kids' Danger?" February 6, 2006. www.cbsnews.com.

Police are investigating possible MySpace connections in both murder cases.

The Center for Missing and Exploited Children reported more than 1,200 incidents last year of adults using the Internet to entice children. With numbers like that, you'd think parents would be hovering over their kids, wanting to know what they're doing online. But authorities say many parents are clueless about their kids' MySpace profiles.

Do [kids] think, "Hey, there's 35-year-old or 45-year-old guys out there looking at my site?"

CBS News Technology Analyst Larry Magid had a look at one personal profile on the site, belonging to a 15-year-old girl.

Magid says the girl writes in her description, "Drink a 40, smoke a bowl, sex is good, life is great, we are the class of 2008."

"Now if you were a predator and you read something like that," asks Hughes, "what would it tell you about this young lady?"

"I'd target her, I think," Magid replies.

Kids Are Not Careful

In talking to some teens who regularly use MySpace, it's easy to see that a lot of kids aren't very careful about the information they put on their pages.

"So many people don't even use common sense," says Katie Pirtle, a high school student. "Some people even put their phone number on there."

And while the information kids put on MySpace may be intended for their friends, do they think, "Hey, there's 35-year-old or 45-year-old guys out there looking at my site?"

"Definitely not," says April Ehrlich, another high school student. "When they think MySpace, they think other teenagers. They don't think there are adults pretending to be teenagers on there."

A lot of MySpace users post "the survey," which asks for responses about issues like drinking, drug use and skinny dipping. Users can also put up pictures.

MySpace declined CBS News' request for an interview. The site warns users not to post any "personally identifiable" material—but the teens we spoke to say that advice is routinely ignored.

"Just like a car accident, it can happen to you," says high school student Julia Rinaldi. "Predators can come to you—and that's what they don't think when they post those things."

Those predators include men like 26-year-old Jeffrey Neil Peters, who was arrested last month for sexually assaulting Susie Granger's daughter. Granger says parents should keep their kids off the site.

"Please don't allow your children to go onto MySpace," she says. "It's a very unsafe environment for them to be in."

But for the thousands of teens who are hooked on the site, it's a warning that's lost in cyberspace.

Pornographers Expose Children to Obscenity at Online Social Networking Sites

Tom Rodgers

Tom Rodgers is a retired detective lieutenant for the Indianapolis Police Department. He now serves as a consultant for law enforcement agencies and not-for-profit organizations on issues concerning obscenity and sexual abuse of children.

In its "To Catch a Predator" series, NBC TV's *Dateline* teamed up with an Internet watchdog organization called Perverted Justice to fight against Internet sex predators. Early episodes of the series showcased the popular website MySpace.com, showing how pedophiles are using this site to shop for children on the Internet for sexual purposes. In these episodes, Perverted Justice investigators, working in conjunction with law enforcement, pose as decoys and make contacts through MySpace.com. As a result, dozens of men have been arrested for a variety of charges.

What is equally disturbing is the large number of On-line Obscenity Complaints about MySpace.com that have been received at Morality in Media's (MIM) www.obscenitycrimes.org web site. Unlike the Dateline series, the MIM complaints are not about sexual predators stalking children online, but about the pornography that is readily available on or through My-Space.

MIM has two consultants who validate these complaints. Both consultants are retired law enforcement agents with experience in investigating obscenity crimes. Once these complaints have been validated, the consultants then write and submit detailed reports to the Justice Department for investigation, follow-up and potential prosecution.

MySpace.com is a site where people can find other people with similar interests and post information on blogs regarding a variety of subjects. It is sort of a cyber social club. People can go there for free and research information on all kinds of ordinary subjects like films, music, forums, comedy, classifieds and more. They can also view pornographic pictures and videos, and children are only a few clicks away from being exposed to pornography at MySpace.

[Between] May 2, 2004, [and September 2006] MIM received approximately one hundred and sixty complaints from citizens about MySpace.com. This includes complaints where children have been exposed to unwanted "porn spam" or to pornography while searching for other content and complaints where children have knowingly exhibited pornography on their own web sites.

Many MySpace web pages also have direct links to commercial pornographic web sites that offer free "teasers."

In an email to MIM, one concerned parent asked, "I would like to know how a website such as myspace.com is able to have pornographic material available through its users if the site was designed for children 14 and over.... Some children under the age of 18 have soft to hardcore porn within their pages, often pictures of themselves. Myspace.com appears to be nothing but a great opportunity for children to have complete access to porn. I would like to know if and what the regulations for this site are (sic)."

Curious onlookers were also subjected to unwanted graphic sexual acts and printed information. One complainant stated, "A large number of underage users are posting and reading obscenity and pictures." A second complainant commented, "16 year old girls allowed to discuss bestiality." A third stated that he/she had received uninvited pornography through an instant message from some unknown person signed on through MySpace. Many MySpace web pages also have direct links to commercial pornographic web sites that offer free "teasers" (pornographic photos or videos that can be viewed without proof of age), in addition to full subscriptions that require payment. Upon checking a recent complaint on MySpace.com, one of the MIM consultants reported that a pornographic video automatically began playing as soon as the MySpace page was opened up.

It would be very difficult to find a young teen that has never heard of MySpace.com. In fact, it is surprising how many teens have actually set up their own MySpace blogs. It is the vogue thing to do. As a teenager, if you do not have your own blog then you have not kept up with the changes. It is shocking to see all the personal and potentially damaging information that people, including teens, are posting about themselves and their families on these sites.

What may seem like an innocent blog site and fun place for teens, MySpace.com can, therefore, be a very dangerous place for children to spend time. So what are the chances of your children visiting MySpace.com? The number of visitors is steadily increasing, as MySpace.com went from 4.9 million U.S. visitors in 2004 (Janet Kornblum, "Teens Hang Out at MySpace," *USA Today*, January 8, 2006) to more than 50 million U.S. visitors in May 2006 (press release, "Social Networking Sites Continue to Attract Record Numbers," comScore Networks, 6/15/06). According to an article in *USA Today* ("The MySpace Invaders," 8/1/06), MySpace had 95 million members and more than 81% of the online social networking traffic.

Pornography is invading just about every aspect of our lives. It has come into our communities with so-called "adult businesses" and into the so-called family video stores that contain the "adult sections." It has invaded our schools, libraries, churches and job sites via the Internet. Because pornography might seem to be everywhere, there are a lot of people who believe that pornography is okay, acceptable, and perhaps even legal. But, as Roger Young, a retired FBI Agent and MIM Consultant, teaches, "Americans have been misled, misinformed, and wrongly educated that pornography is okay, acceptable and perhaps even legal, when it is not!"

In the 1973 *Miller v. California* case, the Supreme Court held that the First Amendment does not protect obscene materials and that obscenity laws can be enforced against hardcore pornography. In 1996, Congress updated federal obscenity laws to clarify that distribution of obscenity on the Internet is a crime; and at this point [2006] federal obscenity laws are the only legal weapon available against Internet pornography because the Child Online Protection Act, which is intended to restrict children's access to Internet porn, is still tied up in the courts.

Most people do not want the Internet to become a cesspool of hardcore pornography. They do not want children being exposed to hardcore pornography. My question is this: When will Americans (parents in particular) get fed up with the flood of Internet pornography pouring into their communities and homes and insist that Internet obscenity laws be enforced?

Gang Members and Wannabes Meet on Social Networking Web Sites

Katherine Leal Unmuth

Katherine Leal Unmuth is a staff writer for the Dallas Morning News.

Teen gang members and wannabes are finding a platform in cyberspace to brag, bully and blog about their exploits.

Social networking sites feature Dallas-area youths dressed up gangster-style and flashing gang signs, holding guns and knives, getting tattooed or partying with marijuana. Common images include pit bulls, gang pledges and rivals' colored flags on fire.

Many of the Web postings come from kids who merely want to imitate the gangster image, which is not illegal, investigators say. But they add that they have to take the sites seriously because some of the postings come from serious gang members and many of the "wannabes" can become real dangers.

Irving's Nimitz High School Resource Officer Ken Richardson says parents need to take the words and photos posted by their children seriously, too. "A lot of times they're shocked," he said of the images he's shown to parents. "To me if they're already acting out, taking pictures and throwing gang signs, the parents should be looking at that and talking with their kids."

Gang-related postings come from teens all over, with screen names such as "oakcliff-crip" in Dallas and "blood-thuglife213" in Grand Prairie. The sites include MySpace and MiGente.com as well as a new site, Bebo.com.

Katherine Leal Unmuth, "Teen Gangs Find Voice on Web," *Dallas Morning News*, June 1, 2006. © 2006 *The Dallas Morning News*. Reproduced by permission.

Bebo officials say membership among schools spreads through word of mouth and note that Irving high school students are "very active," with 1,200 to 1,500 users per school. That may be because the school district provides every high school student with a laptop and wireless Internet access.

Most of the postings on Bebo are not gang-themed. But a feature on the home page for each school allows users to design online graffiti bricks on what's called "The Wall." In Irving, the bricks include prominent references to the mostly Hispanic gangs "South Side 13" in blue scrawl or "West Side 12" in red. Screen names include "lil salvatruchol3," for the infamous Salvadoran gang MS-13; "2800bloc," referring to a block of Pioneer Drive where police have noted gang activity; "surcrips;" and "wssidegangster."

These cyber-gangsters conduct online polls asking what color or gang teens represent. They arrange fights in person and rehash past victories.

One young user, who calls himself Jose, pledges his allegiance to the Sureños 13 gang: click click bang bang. In a blurry photo he stands holding what appears to be a rifle in front of a Mexican flag. "I represent that southside," he writes. "So don't run up and try all that wack [expletive] cuz I will shoot . . ."

"Get strapped quik!!!" another teen writes. "I aint playin!!!"

Several teens responded to a reporter's inquiry about their postings with profanity and threats. One wrote "duh we in a gang," before bragging he was in an FBI book of pictures of West Side 12 gang members.

Police say the sites help them keep tabs on gangs: which ones are active, details about recent incidents, new trends in dress and language.

"I do this cus im proud of my family, my friends and mostly the hood," wrote a 15-year-old who goes by the screen name "wstwix." "it's a promise that u keep 4 the futer generation how to protect ur neighbors show every body that were not punks that we are willing to die 4 la raza u know take a bullet 4 each other."

Monitoring Use

Bebo vice president Jim Scheinman said the site may remove inappropriate content if a member complains about it. The company also is adding a chief safety officer.

For its part, the Irving school district already blocks access to social networking Web sites. But students get around the filters and are often online during school hours.

Superintendent Jack Singley said such activity is an "inappropriate use" of the laptops that the district provides. But he also noted that it's difficult to tell whether students are using their school computers or home computers to post.

Students who violate the district's policy on acceptable laptop use may be punished or have their laptops erased and restored, which costs them $15.

Officials say they may also ban cyber-bullying—threatening or harassing others online—their student code of conduct. "It's a growth and maturity issue, not necessarily a public education issue," Mr. Singley said. He and other educators call for more parental supervision. Maria de la luz Flores, whose children attend Nimitz High, said Latino parents may be at a disadvantage because of language barriers or a lack of education. "There are many parents who know very little about computers," said Ms. Flores, a Mexican immigrant who teaches a computer class to adults. "We need more classes so parents will benefit from the information."

Wesley Fryer, who blogs about technology in education from his office at Texas Tech University, said school districts can't simply block Web sites out of fear—they must foster discussion on appropriate use. He compares it to avoiding talking about safe sex because of worries the information will encourage sex. "Right now we see most school districts banning these sites but not addressing how they should be dealt with," he said. "This is a conversation everybody should be having."

Police Tool

Irving and Dallas police say the sites help them keep tabs on gangs: which ones are active, details about recent incidents, new trends in dress and language. Some members even post photos of themselves creating graffiti.

"Everybody wants to act like a thug or a gangster," said Sgt. Mark Langford of the Dallas police gang unit, which mostly monitors MySpace.com. "But in the middle of all that, there are nuggets of truth."

Irving gang investigator Brett Burkett said he saves students' online profiles, just in case. "During an interview they may say, 'No way I'm in a gang,' and then we show it [their profile] to them," he said.

Precisely what all the Web traffic says about gangs in Irving, however, is a subject of some debate. Irving police say they have no specific numbers on gang-related arrests or assaults and haven't seen other evidence of a growing threat. "While we have gang activity, it's not reached a level that our officers are looking at an increase or decrease," Irving Police Chief Larry Boyd said. But Omar Jahwar of the gang intervention program Vision Regeneration said he's met with gang members in Irving. He said that even if the Web sites are full of wannabes, they often precede the "real deal." "On a civic level, I don't think they're ready to admit how severe their gang problem is," he said. "You've got this urban ideal matched with suburban money. You've got the Las Colinas business community side of the world saying we are not Dallas and the kids saying we are."

Bullies Use Social Networking Web Sites to Terrorize Other Teens

Steven Swinford

Steven Swinford writes for the London, U.K., Times.

Siannii Roughley was sitting at a friend's computer after school in January when she came across the website. At first it looked like any other teenage girl's home page, decorated with a pink background, love hearts and glitter.

As she scrolled down the page Siannii, 13, was shocked to see a picture of herself and the caption: "Well, dis iz da slag Siannii athat no one loikes coz shes a dirty greebo. Plz sign da shout box of wt u think of her xx SHES A SLAGxx." Beneath it were dozens of messages, some telling her to "f*** off and die", while others said: "We're going to get you at school tomorrow." The page had been viewed more than 2,000 times.

"It was awful," said Siannii, from Famborough in Hampshire. "I was terrified to go to school. I wasn't sleeping. I began to think that maybe they were right: maybe I should just kill myself."

The boom in social networking sites has been accompanied by the spread of "cyberbullying," a trend that some experts believe is fast getting out of hand.

A Boom in Social Networking Sites

Siannii's is far from a lone case. The past two years have seen an explosion in "social networking" websites where teenagers meet and chat.

The sector is dominated by two players, Myspace and Bebo, which between them boast more than 100m [million] users worldwide. Myspace, recently acquired by News Corporation, owner of the *Sunday Times*, is by far the bigger and has attracted controversy over child safety issues in the US.

Bebo, however, which was started a year ago, is now level-pegging in the UK with 4m [million] users, and organises its site around individual schools.

Once logged in, teenagers create "home pages" for themselves—profiles in which they detail their lives, loves, friends and other teen obsessions. Most home pages come complete with pictures, daily blogs and even video clips. Once a new home page is posted, others are able to peruse it and post messages on it largely as they choose.

Gradually each child builds up a bigger and bigger network of online contacts and friends—all well away from the prying eyes of teachers and parents.

In a recent survey of 500 teenagers carried out by Microsoft, more than one in 10 said they had been bullied online.

Although largely harmless, the boom in social networking sites has been accompanied by the spread of "cyberbullying", a trend that some experts believe is fast getting out of hand and may be being exaggerated in the UK by Bebo's focus on schools.

In a recent survey of 500 teenagers carried out by Microsoft, more than one in 10 said they had been bullied online.

Stephanie Godwin, a 13-year-old from Gloucester, was subjected to a year of bullying on the Internet. She was sent a stream of abuse on her website and became so scared she refused to go to school. One of the messages read: "Gay! Ur a f****** fat cow and a fat f***. U wanna shut ur f****** mouth cos you is gonna get banged".

Debra, her mother, said: "I couldn't believe what these kids were doing, she just couldn't escape it."

Online Bullying Is of Increasing Concern

For children's charities and schools, online bullying has become an area of increasing concern. In Norfolk the county council has banned access to Bebo in every school, while in other areas head teachers have spent thousands of pounds on monitoring systems.

Children say things that are much more extreme and vindictive than they would in their everyday lives.

Chris Cloke, head of child protection at the NSPCC, said: "It's a completely unregulated world out there. I'd compare it to a modern-day *Lord of the Flies*. Children say things that are much more extreme and vindictive than they would in their everyday lives."

Others welcome the new websites, arguing that the bullying that occurs on them has simply been transferred from the school gates and would happen anyway.

Dr Rachel O'Connell, head of the internet research unit at Central Lancashire University, hailed the boom of social networking as a "cultural revolution". "The possibilities are extraordinary. It's a very positive thing for children and a fantastic creative outlet. There are concerns about cyber-bullying, but for most children this is a very exciting and immersive technology."

Who is right? Are social networking sites exacerbating bullying or merely mirroring the cut and thrust of the school playground?

Parents thinking of sneaking a peep into one of the new sites should prepare themselves for a shock: it's like entering a parallel universe.

Teenage girls fill their pastel pink and blue home pages with provocative pictures of themselves and schoolyard gossip, while boys boast about drinking feats (real and imagined) on pages festooned with pneumatic babes and cars. Evryl tlks n txt, init. And all—almost without exception—are obsessed with sex.

"I'm sexy, I'm cute! I'm popular to boot! I'm bitchin', great hair! The boys all love to stare," announces Sarie, aged 14, in London.

Bullies have taken to setting up fake home pages in their victims' names in a bid to humiliate them publicly.

Tom, 13, from Hull, is similarly confident: "I have blond hair blue eyes and have the capability 2 sweep any girl off there feet uless there stuck up sooo check me out sexy!!"

A Window on the Youth Subculture

According to Sonia Livingstone, a professor in social psychology at the London School of Economics, the sites provide a "window on youth culture". "Most of what they're doing is fun and innocuous: it's teenagers being teenagers," she said. "It can also provide a space for them to discuss things they couldn't otherwise, like their sexuality or health issues."

Dig beneath the surface, however, and a darker picture of cyber-bullying emerges, with some of Britain's most prestigious public [i.e., private]—schools leading the way. At Tonbridge school in Kent, bullies have taken to setting up fake home pages in their victims' names in a bid to humiliate them publicly. One accuses a boy, whose name and photograph appear on the page, of incest and bestiality.

Another attacks a named pupil for being overweight. The forged profile says: "I am a fat shit and wish i had some friends who liked me. I am 6ft tall and 6ft wide. I weigh the same a double decker bus. I am proud of my weight because i created third world hunger."

Ray Hart, the school's bursar, said last week that the matter would be investigated and insisted that the school was taking measures to clamp down on e-bullying.

Any abuse is broadcast to the whole world and leaves a permanent record.

At Dulwich college, south London, it is not just the pupils who are victims of online bullying. Students have also created fake profiles of two teachers at the school. One announces next to a picture of the master concerned: "I am just a bit oooooh, you know? I love my toy boy, he is fit! ... my hobbies are reading (preferably male porn), smelling roses, dressing up as a woooooooooman, swearing, watching rugby especially scrums and drinking milk straight from a cow."

Serious Attacks Have Occurred

Even more serious abuse has taken place in Ireland. In May, Michael McIlveen, 15, was beaten to death in a sectarian gang attack after eating a pizza with a friend from a loyalist estate. After the murder, an innocent teenager's Bebo site was bombarded with abusive messages accusing him of complicity in the murder. One read: "I hope all of you f****** sick bastards die and your bodies are left in the street so we can all walk past and jump on your heads."

However, some of the most vicious cyber-bullying is aimed at girls. In one case described to The Sunday Times, a mother was distraught to find explicit references to her 13-year-old daughter's alleged sex life on Bebo. The victim's boyfriend and his classmates had set up a home page which described her as a "slag" who had indulged in a list of explicitly spelt-out sexual acts. They also sent the girl messages to her own site sarcastically asking her for sex.

The mother discovered the site after her 17-year-old son drew her attention to it. "I find it horrific that they can treat girls in this way. It shows a complete lack of respect. This has upset the whole family," she said.

Such cases have led some schools to ban Bebo entirely, and others to suspend and even expel pupils.

John Horsfield, head teacher at Neatherd high school in Dereham, Norfolk, said: "My concern is that children are so unprotected. No matter how much information we give them, they are still giving out their addresses on that site, along with photos of themselves and their houses. We live in such a dodgy world."

The British co-creator of Bebo, Michael Birch, last week fiercely defended his site and others like it, arguing that bullying, although a problem, represented only a tiny fraction of the exchanges between users. "Nobody out there is saying we should close Bebo down. Social networking can be a positive and creative thing," he said. "(But) we have a responsibility both legally and socially to make it as safe an environment as possible."

Not Enough Is Done to Make Sites Safe

Critics counter that sites such as Bebo are not doing enough to make them safe. Nathalie Noret, a lecturer in developmental psychology at York St John University College, has conducted several studies into cyber-bullying. "The anonymity behind cyber-bullying means they can say whatever they want. If they're in the playground saying this, there's the chance a teacher might hear or another pupil might step in. This is an unregulated space," she said.

More significantly, any abuse is broadcast to the whole world and leaves a permanent record. One of the most famous cases involves an Internet video of an overweight 15-year-old having an imaginary light sabre fight. The two-minute clip of Ghyslain Raza, from Quebec, was posted on the internet as a joke by schoolmates, and has since been downloaded more than 12m times. Raza is now suing his classmates for £140,000 [about $272,000] for making him a laughing stock.

O'Connell, who will soon be joining Bebo to help with child safety, said: "If kids bully someone online they create a permanent digital record of their bad behaviour. When you're going for a job and your boss Googles you, he will see that stuff. We need to encourage children to be responsible when they go online."

Bebo relies on its users to report any abuse. According to Birch, it responds quickly to complaints by cancelling memberships and removing offensive home pages. In Britain, dozens of accounts are removed on a daily basis and Birch plans to introduce a system to screen each of the 700,000 images uploaded onto the Bebo site every day and work more closely with schools. In the United States, the company alerted police after finding indecent pictures of children on the site.

Birch also plans to work more closely with schools, enabling them to appoint responsible children to oversee any new users and monitor the site.

He admits that keeping tabs on the content is difficult: "Cyber-bullying is a headache but it's not a simple thing to address, partly due to the sheer volume of activity. Searching for swear words will not do it alone. But you can't do something so negative that it kills the community."

Birch estimates that within two years social networking sites will overtake Google as the most popular on the Internet, with Bebo having a potential 300m [million] users worldwide.

The Problem Will Not Go Away

For Liz Camell, head of the charity Bullying Online, it is an alarming prospect. "The bigger they get, the worse cyber-bullying will become. It's not a problem that is going to go away," she said. "When you're relying on young people to moderate themselves, with the best will in the world they're not going to do that. It's not beyond the bounds of possibility that these sites will get sued."

Some schools are investing in £9,000 [about $17,500] monitoring system to detect offensive messages and pictures on school computers, and to alert teachers.

Colin McKeown, managing director of Zentek Solutions, has installed monitoring systems in 30 schools in the north-west of England and discovered everything from death threats to suicide notes. "Any normal thoughts of how we perceive the bully have changed dramatically," he said. "There's no longer the disaffected big thug in the playground. These are bright, middle-class children as well as working-class youngsters and it escalates very quickly."

Schools Use Social Networking Web Sites to Catch Athletes Who Break Rules

Maria Sacchetti

Maria Sacchetti is a staff writer for the Boston Globe.

High schools across Massachusetts are threatening to punish athletes if they are spotted drinking alcohol or using drugs in photos or videos posted on MySpace, YouTube, or other online sites.

School officials say they are enforcing existing bans on smoking or drinking, and turning to online sites to catch the rule-breakers. In at least 20 high schools across the state, principals are warning athletes that they will punish them for behavior caught online, according to the state's secondary school principals group and athletic association. The two groups estimate that dozens of schools are using this tactic. Several schools have suspended students from games.

Woburn High School suspended a handful of athletes from two practices and one game last spring after police recognized the athletes holding cans of beer in photographs posted on MySpace. This year, Newton South High School notified athletes they could be suspended if captured breaking the rules in photographs or video online.

Schools generally do not punish nonathletes for behavior outside school, but the 175,000 student athletes in Massachusetts must follow a code of conduct that bans drug and alcohol use during the season. The Massachusetts Interscholastic Athletic Association, which governs school sports, requires schools to, at a minimum, suspend first offenders for 25 per-

Maria Sacchetti, "To Catch Rule-Breakers, Schools Look Online," *Boston Globe*, December 22, 2006. Used by permission of *Boston Globe*, conveyed through Copyright Clearance Center, Inc.

cent of the games and subsequent offenders for 60 percent. Individual schools can set tougher rules, including removing students from teams or enforcing the rules year-round.

Online evidence may be questionable legally, said lawyers who work on civil rights and education issues. Photographs online could have been doctored.

Schools already have been warning students to be careful about what they post online, but punishing athletes for misdeeds online is a more aggressive approach.

"We're dealing with a new world," said Ron Lanham, Newton South High's athletic director, who said he would rely on tips and doesn't plan to comb through the sites. "We don't want to be seen as the MySpace police. But we're looking out for the best interests of our athletes."

After Newton South officials spoke with students about the new policy, some expressed surprise, saying they thought the information they posted was private. School officials informed them that the Internet is as public as the town common.

But online evidence may be questionable legally, said lawyers who work on civil rights and education issues. Photographs online could have been doctored, and the date they were posted could be unclear, said Sarah Wunsch, a staff lawyer with the American Civil Liberties Union of Massachusetts.

Schools Should Ensure That Students Get a Fair Hearing

Wunsch said schools should ensure that students get a fair hearing, which the athletic association's rules require.

"There may be some due process rights for kids about the fairness of this," she said. "Anybody could use Photoshop and stick somebody's head on that; you could get somebody in trouble."

The courts have upheld an athletic association's right to set special rules for athletes because sports are not required in school, said Paul Wetzel, spokesman for the Massachusetts athletic association.

Principals said they tread carefully once they get a tip about content on a website. Bob Norton, Woburn High principal, said he investigates student behavior only when he gets a police tip about a student breaking the rules. Last spring, he said, a police officer conducting an unrelated investigation recognized some of the town's athletes in photographs on MySpace that showed "clearly a party situation." Police forwarded the photos to Norton. He said he interviewed the athletes and they confessed.

"I don't sit here in the afternoon by myself and say, 'Let's see who I can catch on MySpace,'" Norton said. "What we are trying to do is have our kids do things that are healthy and safe. I'm not interested in the penalties as much as I am interested in the message."

At Newton South High, several students recently told the school newspaper, *The Lion's Roar*, that they supported the policy. In a Globe interview yesterday, junior Courtney Chaloff, captain of the volleyball and cross-country ski teams, said she hoped the policy would deter students from drinking or using drugs. She said she has seen pictures of students at her school with beer bottles online.

"I don't think athletes should be doing that sort of thing to their bodies," said Chaloff, 16, who frequents Facebook, another website. "I would never do it, and if I did, I would never post pictures of myself online."

Her mother, Jamie Chaloff, said she supports the school's policy. "They're really only doing it for the good of the kid," she said. "There are so many parents who don't know what their children are doing."

School and athletic association officials said they expect enforcement to increase as school officials become more Internet-savvy.

Robert Gay, North Attleboro High's principal, recently informed all student athletes that they could be punished for behavior caught online.

"I have no doubt that this is going to be an issue," said Gay, also president of the Massachusetts Secondary School Administrators' Association. "If the kids are foolish enough to post a photograph, or someone posts a photograph that they're in, most principals would take some sort of action on it."

Authors Use Social Networking Web Sites to Connect with Readers

Rachel Kramer Bussel

Rachel Kramer Bussel is a writer and editor.

Though MySpace bills itself as "a place for friends," the networking site has become much more than just simple friendship: These days, it means business, and not just for owner Rupert Murdoch. First, musicians started flocking to the site as a way to promote their new releases and gain exposure by converting MySpace friends into listeners and fans. Now getting in on the act are authors, who by banding together with other writers, racking up thousands of friends interested in their titles, setting up pages for their characters, and holding online contests for writers and readers both, are using MySpace to garner their books a broader readership and help their publishing careers flourish. . . .

MySpace's recently-launched Books area currently ranks 551 books by popularity (read: number of MySpace blog links) under "Top Books"—at the time of this article's publication, *Danse Macabre (Anita Blake Vampire Hunter)* by Laurell K. Hamilton and Lauren Weisberger's *The Devil Wears Prada* were tied at first place with 32 links each. Titles ranked on MySpace link to Amazon (via its affiliate program), while the Books area also boasts a featured book blog, highlights three genre-based book groups, and showcases five featured titles—with unattributed reviews, as well as reader comments—on the Books page. Since MySpace marketing and content executives declined to comment for this story, it's unclear how featured books are chosen, or how MySpace plans to further develop its Books section.

Clicking with Readers

A primary reason authors join MySpace is to connect easily—and instantly—with their existing audience, as well as thousands of potential readers. Elsewhere on the Internet, many writers employ Web designers to help maintain personal sites, making it costly and time-consuming to keep them freshly updated, especially with last-minute news . . . [one author who], kept a MySpace blog while on her book tour, and says the "control freak" in her likes being able to notify her fans with the touch of a button when she's doing a reading or has a TV appearance. Zailckas says she feels closer to those she meets on MySpace, since she can check out their profiles just as they view hers. "It's more intimate," she says. "The readers you're communicating with are honest-to-god human beings with faces and (now and again) your book in their list of 'favorites.' Silly as it sounds, there's something tremendously moving about being able to 'see' one another, even if it is in a nerdy sort of way." Zailckas points out that while readers aren't necessarily visiting their favorite author's Web page religiously, they often log on to MySpace on a daily basis and will see bulletins posted there.

> *It's intimidating to write a fan letter to an author. . . . But in the MySpace universe, somehow that kind of inhibition and difficulty has been broken down.*

Author Elisha Miranda has been using MySpace to try some "guerrilla marketing techniques" to push her first novel, *The Sista Hood*, a young adult title geared toward communities of color. Her hook? A CD featuring original lyrics inspired by the novel, and MySpace pages for each character, which lets readers track characters' lives beyond the book's conclusion. Miranda relies on MySpace to help her better understand where her audience is coming from. "There will be a few young women who have very 'hoochified' sites," she says, "but when I take the time to get to know them, there is so much more there. They want to talk and—yes—read. What a missed opportunity if I allowed my judgments as a feminist to get in the way."

As they describe supporting their books as one might devote oneself to a campaign for social change, many authors say a chief reason to use MySpace is to "raise awareness" for their titles. Brad Listi has been blogging every day on his MySpace page to promote his debut novel *Attention. Deficit. Disorder.* For Listi—who admits he "co-opted" the MySpace Music template prior to the launch of MySpace Books so he could list his book tour dates—the site helps his prospective readers get to know who he is. "It allows writers to introduce themselves to readers in a manner that exceeds the standard advertising solicitations and naked commercial backflips," explains Listi. "What I do doesn't involve any kind of hard sell. It involves taking the time to write each day, answering all correspondences with readers, and building a fan base in a personal, organic, and entertaining way." . . .

MySpace Friends . . . with Benefits

Marcy Dermansky has over 3,000 friends on MySpace, and has used her Top 8—the eight hand-picked "friends" MySpace users see listed when they visit someone's page—to promote her novel *Twins.* All eight of Dermansky's featured friends share names of characters in her book, including Sue, Chloe, Yumiko, and Smita. To find MySpace friends with the more unusual names, Dermansky did searches and sent emails, finding some recipients so responsive they wound up buying her book and wanting more information on their fictional namesakes. By creating a presence for the book on MySpace, Dermansky says she feels "certain that it has increased sales and raised my profile as an author." She enjoys the give-and-take with readers, and feels that by simply logging into the site on a regular basis, she presents herself as a more accessible author. "It's intimidating to write a fan letter to an author. There is the fear of being dorky or inarticulate. But in the MySpace universe, somehow that kind of inhibition and difficulty has been broken down."

Taking Dermansky's idea and going a step further, many authors set up full-blown pages for their books' characters. Karyn Bosnak's done that for her *20 Times a Lady* protagonist Delilah Darling, but isn't sure yet how she'll tackle it. Her original plan was to make pages for all 21 characters in the book but she's since scaled down to focus on just Delilah. Bosnak doesn't just appreciate the author/fan interaction MySpace can foster, but also uses it to research her readers' book preferences. "I'm a visual person—MySpace gives me a chance to put a face on the emails I get, and provides an opportunity to see what else my readers are reading."

While they can't quantify whether MySpace has had a direct impact on sales, many authors say it's definitely helped them boost event turnout. Christina Amini and Rachel Hutton, editors of the anthology *Before the Mortgage*, even used MySpace to draw one of their favorite actors to their book. "We checked out the page for BJ Novak, who writes for and plays the temp on *The Office*. After finding out [via MySpace] that he loves homemade Chex Mix, we emailed him to say we'd make his favorite treat if he came to our L.A. reading. He did, and ended up enjoying the book as much as our Chex Mix."

Teen literature is a natural growth area on youth-oriented MySpace.

Horror author Michael Laimo reports that hundreds of MySpace fans have told him they've purchased his books, saying he's gotten over a dozen interviews through his page. "I've also made my first movie deal through MySpace, after an independent director contacted me about film rights." Author

Josh Kilmer-Purcell actively courts readers of David Sedaris or Augusten Burroughs, whose writing style resembles his own. "When I go on a rampage adding [MySpace] friends, my Amazon ranking will increase noticeably for the next few days. But it's not about selling—it's about awareness," he says. "There's not a lot of mass media devoted to books, so when I find someone who likes authors similar to me, I want to make them aware that I exist."

Remapping the Writers' Group

While most authors have adapted the basic MySpace page to suit their needs, some have gone in a whole new direction, forming groups to bond with and promote their fellow authors. Teen literature is a natural growth area on youth-oriented MySpace, and Sarah Mlynowski, who writes both chick lit for adults and teen lit, has formed a Teen Lit group there, numbering more than 600 members, over 60 of whom are authors. The space offers participating readers and aspiring writers of the genre a chance to ask questions about everything from PR and MFAs to school bullies. There, group members network and post book and tour information, aimed at creating a sense of community and referring readers interested in one author-member's work to similar books in the genre they might also enjoy. [As of March 2007, this group had over 1800 members, over 100 of whom are authors.]

Going a step further is MySpace's Memoirists Collective, formed by authors Hillary Carlip (*Queen of the Oddballs*), Maria Dahvana Headley (*The Year of Yes*), Josh Kilmer-Purcell (*I Am Not Myself These Days*), and Danielle Trussoni (*Falling Through the Earth*). Though they work with different publishers, they've united in their MySpace group to promote their recent books with contests and feedback, and jointly encourage aspiring memoirists. . . . Kilmer-Purcell cites the sense of camaraderie MySpace fosters amongst his fellow memoirists, saying, "I get tired of promoting in a vacuum. We're not competing—the stakes aren't worth it." As Headley puts it, "Our books are all different, and one of the cool things about the Collective is that our readers are cross-pollinating. We have everything from dominatrices to soccer moms on our friend list. Some of the people who are hearing about my book [on MySpace] would never have crossed my path any other way."

Political Candidates and Nonprofit Organizations Use Social Networking Web Sites

Kate Kaye

Kate Kaye is an editor at the ClickZ *Network, a Web source of interactive marketing news, opinion, and research.*

When it comes to politics and social media, the "C" in CGM [consumer generated media] stands for "citizen," not "consumer."

But in many other respects, the political campaigners and issues advocates who are dogging the MySpace crowd have a lot in common with commercial marketers.

Like their profit-driven brethren, political organizations and nonprofits are trying to determine how best to integrate CGM into campaigns. And like those marketers, the political set hopes to find the right balance between encouraging supporters and maintaining control over campaign messages.

More and more campaigns are starting to think about how they can use new digital communication outlets like blogs, RSS feeds and social networking sites.

"The challenge as a political candidate is you don't want to dampen anybody's enthusiasm, but you don't want someone running your campaign who's not you," explained Mark SooHoo, VP of Campaign Solutions, a political consulting firm that works with right-wing candidates.

More and more campaigns are starting to think about how they can use new digital communication outlets like

Kate Kaye, "Political Campaigns and Nonprofits Sort Out CGM," *ClickZ Network*, June 19, 2006. Copyright © 2006 Incisive Interactive Marketing LLC. Reproduced by permission.

blogs, RSS feeds and social networking sites. Observed SooHoo: "Because of all those things together they're beginning to invest more in their online strategies and take a comprehensive look at all of these things."

Online social networking and CGM have become integral parts of PETA's [People for the Ethnical Treatment of Animals] marketing strategy, for instance, specifically when it comes to reaching out to teens. The animal rights activist group posts video to YouTube, IFILM, MySpace and Google Video. "These are videos that wouldn't be approved for TV because of graphic content," said PETA Marketing Manager Joel Bartlett.

Bartlett soon may have more assistance in managing relationships with netizens; He recently got approval to hire an e-community coordinator. The new staffer will be responsible for overseeing the group's online message boards, podcasts, blog, and mobile marketing campaigns, as well as coordinating its efforts with those of supporters who upload images and video to sites like Flickr and YouTube. PETA already posts videos submitted by users to its PETA TV site, and sends out daily bulletins to over 64,000 MySpace friends.

"Historically, PETA was intimidated by giving more control to the users," recalled Bartlett. Those days are long gone now that PETA has built a Street Team network of over 160,000 teens over the past four years, many of whom use online message boards, MySpace pages, and other Web venues to get the word out on animal rights issues and demonstrations.

"We believe that we need to put an effort into getting people to do [CGM] more and more," said Bartlett.

Letting the Netroots Grow Wild

Some campaigns are more hands-off when it comes to CGM-spawning advocates. Consider the "Nedheads," a YouTube group created by supporters of Ned Lamont, a Democratic Senatorial Primary contender whom many believe is gaining ground because of his netroots backing.

"They know they can't control it," suggested YouTube user Scarce, the Lamont fan who initiated the group on the free video distribution site. Although Scarce said he has talked to people on the Lamont campaign, Lamont's staff hasn't gotten too involved with the Nedheads' grassroots video collective, which sprung up on its own, independent of the campaign.

Scarce, who declined to share his full name, spends hours each week joining in online discussions, commenting on blogs and creating Web video clips in support of the candidate, who's running against incumbent Senator Joe Lieberman in Connecticut's August primary. Along with a core group of a couple dozen like-minded activists, the amateur video producer posts pro-Lamont speeches, TV appearances and other footage to YouTube. The Nedheads have posted over 125 clips to the site since their group of nearly 500 people was established in late March. The Lamont campaign could not be reached by publication time.

"This kind of stuff is just beginning," observed Scarce. "People are a lot more aware of what the Web can do for them, and Joe Lieberman is aware of what it can do against him." Scarce may be right. Search YouTube videos and groups for Joe Lieberman and you'll find over a hundred videos, many critical of the Senator.

The Lamont campaign is "a real people-powered campaign," said Michael Bassik, VP Internet advertising at political consulting firm MSHC Partners, who contends that underdog efforts like Lamont's tend to benefit from less top-down management, whereas a more established candidate's campaign may require more message control to avoid citizen-generated-headaches.

My(Campaign)Space

A search for "Phil Angelides" on YouTube results in some 20 videos comprised of TV ads produced by the campaign, television appearances and footage shot at campaign events. But the recent California Democratic Gubernatorial Primary winner has his share of detractors. One video post entitled "Phil Angelides Cash" scrolls through a long list of donors from outside California. Another plays a negative TV ad for Steve Westly, his failed primary opponent. Both politicos have My-Space pages.

A post to Angelides's MySpace blog from a campaign volunteer last week prods supporters to help boost the candidate's MySpace friends tally, requesting that supporters post about Angelides as well as rank him in their top eight friends, so his picture appears on their main profile pages.

"No one really knows if [CGM] is going to have a persuasive impact but that's not necessarily the point," concluded MSHC's Bassik. He stressed the fact that new online outlets enable supporters to build communities and meet other like-minded folks.

Indeed, all these new-fangled methods of communication can be considered digital extensions of traditional campaigning. Campaign Solutions's SooHoo put it simply: "All of politics is social networking. . . . Really, it's just going out and saying, 'Hi, I need your vote.'"

Perhaps the most notable example of an organically-grown political Web community was the one that fueled Howard Dean's Democratic Presidential Primary run in 2003. Campaign manager Joe Trippi has been quoted as saying that he hadn't realized the potential of the online community site Meetup until he saw the throngs of Dean supporters who showed up at a Dean Meetup event in New York.

Edwards Dives In

Though some campaigns have yet to incorporate social networking and CGM into their strategies, Democratic Senator John Edwards has embraced them for about a year now. He held a dinner for bloggers in his home last June, and did his first podcast about a year ago, in conjunction with The One

America Committee, an anti-poverty organization he runs with his family. The group also posts videos on YouTube. "We post our video on YouTube because there's an audience there," commented One America Committee Technology Advisor Ryan Montoya, who believes the YouTube posts are helping attract an audience that wouldn't otherwise be engaged in political issues.

Now Hiring CGM Master, Apply Within

One America members have taken initiative to add their own touches to the organization's Web site. They've created a diary featuring a weekly roundup of Edwards-related news, which according to Montoya, has become one of the most popular diaries on the site. "That's something that we can't do, but the community can do it and they do it quite well," he said, acknowledging a need for more help in the CGM department.

"There is a need. It's one thing to keep a Web site up to date; it's one thing to keep a blog up to date," he explained. However, whereas campaigns once only needed a Webmaster, "Now you need a CGM master." Montoya continued, "There are new roles and there will be new roles created in several organizations to adapt to this stuff."

Though most major senatorial, congressional and gubernatorial campaigns have someone on staff dedicated to interactive communications tasks, most have no one who deals specifically with social networks and CGM, according to MSHC's Bassik.

"I don't think a whole lot of people are changing their [campaign staff] structures just yet," commented SooHoo.

However, with devoted volunteers like Nedhead Scarce, they may not need to. "We're going to keep on doing what we're doing," promised the Lamont advocate.

Online Social Networking Is Being Used to Oppose Genocide

Ivan Boothe

Ivan Boothe is the director of communications for the Genocide Intervention Network.

The Genocide Intervention Network and its partner organization, STAND: A Student Anti-Genocide Coalition, both began their life as on-campus student organizations. Now partners and leaders in the anti-genocide movement, STAND encompasses student groups at more than 300 colleges and 200 high schools, while GI-Net provides effective tools for all individuals to help stop genocide through advocacy and fundraising for civilian protection.

How did these organizations go from small student groups to national non-profits in less than two years? Among other methods, we reached out through online social networking websites—sites that promote connections and collaboration between people who share similar interests, geographical backgrounds or schools.

Our interest in social networking started mostly out of necessity. Students and young people have been at the forefront of the anti-genocide movement, and they also have been some of the earliest adopters of new online technology. As a lot of students were self-organizing using MySpace, Facebook and other places, using these social networking tools was a matter of catching up, going where the young people already are, and connecting them to each other.

Ivan Boothe, "Using Social Networking to Stop Genocide," *idealware*, October 23, 2006. www.idealware.org. Reproduced by permission.

Yet even as the organizations have grown—and, for GI-Net, expanded well beyond its student origins—social networking has proven to be an important way to develop and engage our membership.

Our long-term approach to building the movement . . . coincides quite effectively with many of the benefits of social networking: education, collaboration and collective action.

While our current focus is on Darfur, our larger goal is to create a permanent anti-genocide constituency so that when mass atrocities occur again—as they likely will—the world will be well-positioned to respond quickly and effectively. The primary reason countries take so little action to stop genocide is because there is no political will to do so. Our long-term approach to building the movement thus coincides quite effectively with many of the benefits of social networking: education, collaboration and collective action.

While STAND and GI-Net do some traditional fundraising and lobbying, our primary focus is on empowering our members. So, for instance, most of the donations we receive go to fund civilian protection in Darfur, rather than our efforts alone—meaning our members have a real, concrete effect on the ground in Sudan. Similarly, our advocacy is often focused on giving our members effective tools to engage in their own actions, such as our recently-released Darfur scorecard. With a social networking approach, we can extend this empowerment approach to people who otherwise would likely have not come in contact with us.

Using Community Profiles

So what specifically are we doing with social networking tools? We primarily rely on Facebook and MySpace, two of the most heavily-used websites among high school and college students (in fact, MySpace is the sixth most-visited website on the Internet). We set up a group profile in both tools, which is fairly straightforward.

On our profiles, we publish the latest news from our organization and links to current campaigns. We use Facebook's messaging service and MySpace's blog and bulletin features to adapt our action alerts into slightly shorter, punchier versions that are then seen by all of our "friends."

Facebook does not allow much customization of the profile beyond a logo and description, but with MySpace, you can really go wild. There are several free services (such as MyGen, LayoutsPlus, and Zobster) that will help you customize your profile without needing to do too much direct coding. Japhet Els, an online organizer for the Rainforest Action Network, recommends that your profile be "hip"—"If they don't really love your homepage they probably are looking for something a bit more edgy," she writes. For some examples, see the MySpace Profiles for GI-Net, Oxfam, Military Free Zone, or the Marine Corps, or GI-Net's Facebook profile.

We've found that attracting friends is easier than you might think—though it might consume a fair amount of time. If you have demographic information for your members, you could contact those under a certain age once you have your profiles set up. One strategy that we employed was to look for other organizations or ad-hoc groups, and then invite their members. Unfortunately, neither MySpace nor Facebook allows you to do this all at once—you'll have to click on each individual to invite them, one at a time. When STAND launched their national profile on Facebook, they chose instead to send a message to each of the administrators of the local STAND profiles—which had sprung up organically already—and asked them to send a message to their users.

We also post some of our action alerts to related groups in LiveJournal and have plans for expanding to Friendster, but haven't had the resources to undertake that at this time.

We have used social networking sites for advertisements as well as profiles—this was one way we're able to attract more than 400 students to DC for a conference and lobby day just a few months after we were formed. Similarly, a targeted campaign of Facebook ads in Indiana netted us a large number of students willing to call Sen. Richard Lugar's top donors (which we collected from opensecrets.org) and ask them to pressure the senator to approve a bill on Darfur he was holding in his committee.

Integrating Multimedia

GI-Net has, to a lesser extent, used both images and video to spread our message. We have a paid Flickr account on which we post images from Darfur-related events. Our members who have Flickr can post images in their own accounts and use the tag "antigenocide," causing them to show up in a photostream of the latest images from the anti-genocide movement. Our newest website, TimeToProtect.org, features this photostream on the home page.

Because GI-Net and STAND have such small staffs, it has been difficult to fully realize the potential of many aspects of social networking. One small step we took, however, has been pretty successful. We spent a couple of days creating a short video on Darfur using the Apple iMovie software. We posted the video to YouTube—like a moving version of Flickr, and another top-visited site on the Internet—and also posted links to it from our website, MySpace and Facebook profiles. It's hard to measure the direct effect of the video, but soon after we posted it on our MySpace account we started getting 20 or more friend requests each day. That's probably because we made it easy for people to cut-and-paste the code for the video into their own profiles and emails allowing it to quickly spread throughout the site. For another good example of outreach through multimedia, see PETA's Online Action Center.

Extending the Conversation

A final aspect of social networking is the blog community, in which writers post stories online and cross-reference their stories to other websites and other postings. We haven't had time to engage in this as thoroughly as we would like, but we're starting to get our feet wet.

On our primary site, all of our regular "stories"—action alerts, newsletters, weekly news briefings, press releases, etc.— are released through blogging software known as WordPress. This is software one installs on an organization's website, a different option is a hosted service like Blogger. We gather all of our stories together and use the FeedBurner service to publish it as an easily accessible "feed." This makes it easy for bloggers to comment on our stories, as well as the more technically-inclined to subscribe to our live feed in browsers such as Firefox or in specialized news readers.

> *Social networking requires commitment—you can't set up a MySpace profile and then walk away.*

Much of the approach of advocacy in the "blogosphere" draws on a book called The Cluetrain Manifesto, which argues that "markets are conversations" because of the way the Internet has allowed consumers to talk among themselves—and more significantly, to talk back to companies. In recent years, membership and online advocacy has been more thoroughly realized as a conversation, rather than simply a mass of followers paying their dues and repeating the latest talking points.

Despite our constraints of time and resources, GI-Net tries to contribute to this conversation by, for instance, adding to the Wikipedia entry on Darfur. Wikipedia, a collaboratively-edited encyclopedia, is another top-visited site—in fact, the few links on Wikipedia pointing to us generate nearly half of all of our visitors from other sites.

Our website hosting the legislative scorecard on Darfur is built on a content management system known as Drupal, integrated with our Democracy in Action account. One of Drupal's available modules is a "social services links," which you will see on the right-hand side of many pages. These services—del.icio.us, Digg, Magnolia and Technorati—serve a variety of purposes allowing people to bookmark, rate and blog about the pages they link. Although we have spent very little time integrating these services into our campaigns, they will likely become increasingly useful as our commitment to blogging increases.

Supporting the Effort

While none of these approaches cost a lot of money, they certainly require staff time to support. Social networking requires commitment—you can't set up a MySpace profile and then walk away. You have to approve new friend requests, respond to messages, post your latest action alerts, send out bulletins, keep your profile up-to-date, and more. You're not creating a billboard, but rather starting a conversation—you have to be willing to respond.

What does this mean in terms of time? As the director of communications, I spend about 5% devoted to purely social-networking related things (updating the MySpace page, sending out bulletins, adding people to the Facebook group, adding photos to Flickr, etc.) and about 10% on related things (enabling a lot of those features on our sites, like the del.icio.us and Digg links and Flickr gallery). We also have a communications intern and a membership intern who have been helping us since mid-September, but other than that we really don't have anyone committed to social networking. We'd really like to do more things, like have a full-fledged blog and begin talking about these issues more in-depth—not least to engage the blog community more substantially—but we just don't have the staff time to devote to it.

Stepping Back and Letting Go

There's a mentality shift required to fully engage with social networking and community content sites: sometimes, you have to let go. It's true that someone could start posting pictures on Flickr of irrelevant things and tag them "antigenocide," or that some of our biggest fans hosting the YouTube video on MySpace might have profiles that would make some of our donors cringe. Perhaps even more significantly, it's possible that some of the people who encounter us on these sites never make it to our main website, never sign up for a newsletter, never complete an action, and never make a donation.

For us, it goes back to our mission: to empower our members to prevent and stop genocide, and in so doing, to create an educated anti-genocide constituency. While we do, of course, want to increase our membership rolls and make ever-larger donations to civilian protection, in some respects it's not always necessary for people to perform every anti-genocide action through our organization. If our videos or emails or profiles get people talking more substantially about genocide—and the concrete ways in which they can actually prevent and stop genocide—then in some sense whether they end up on our mailing list is somewhat beside the point. Through their knowledge they will engage others, and ultimately enhance the anti-genocide movement we're helping to build. Your goals and your experiences on this front may be different, but to really see the full power of social networking, you have to let it be a network—and in some cases that is going to mean sending something out into the ether, stepping back, and letting go.

Christian Young People Use Social Networking Web Sites

Steve Hewitt

Steve Hewitt is editor-in-chief of Christian Computing *magazine.*

Myspace.com now reports that over 65 million people have set up personal Websites and blogs using their service. AOL and others have announced they are, or soon will be, launching "myspace" types of services. What is Myspace .com? It is many things to many different people. I believe it is evolving into a new way to communicate, and I believe it is something that many in the church need to check out. The good news is that IF it is a reflection of the world (and it appears it is) Christian sites [profile pages] are being created by the thousands by Christian young people who are using My Space.com to share their faith, talk with other Christians, and much more.

If you don't know what MySpace.com is, you can rest assured that your young people do. People of all ages can set up a MySpace.com blog site (both of my sons have sites, and they are in their 20's.) Because it is free, it has become THE place to be if you are a teenager. You can post your picture(s), a video, your favorite music, your personal information (age, sex, etc.) and blog about whatever you desire. If you want to express yourself, gripe about your parents, school or little brother, or if you want a place where you can remain anonymous and ask probing questions about life, MySpace.com is where you will set up a site. People visiting your site can respond with comments, pictures (if you wish to post them) and links. Grouping with other MySpace.com sites is popular.

MySpace.com provides a combination of message board, list serves and Instant Message technology into one site. Coming soon, EarthLink and Korea's SK Telecom have partnered up to offer a new device on the American market from Helio, that will look like a cell phone but more importantly it is a MySpace.com interface device. You can read comments on your site, take a picture and post it to your site, and browse other sites. In this way, you can respond and participate in an ongoing discussion on a MySpace.com site even when you are away from your computer.

MySpace Content Is Varied

The content on MySpace.com is varied. While there are many Christian sites, there are also many sites promoting other religions as well as those that are anti-religious. Many people have talked about the inappropriate materials on MySpace .com, but frankly after spending over 12 hours searching through profiles, I found very little material that was objectionable. However, I was not looking for inappropriate material.

Of course, security concerns exist. Since it is a place for young people to communicate on the Internet, predators will be watching and trying to contact young people through the service. While MySpace.com has an age limit, (you are supposed to be at least 14), they have no way to enforce it and many children are setting up sites claiming to be much older than they are which is dangerous. However, THIS is where the young people of our nation are gathered. This is where they are open and honest, and where (praise God), Christian young people have begun to shine!

In this month's "Internet for Christian News" column, I am only providing information on links to sites [profiles] I found on MySpace.com. I was very excited and inspired to find and browse these and many other sites set up by Christian young people. Many sites have been set up as a place for people to come and discuss things that relate to their Christian walk. For example, one site had a discussion that started when a new Christian shared that they were having trouble during their private pray time with terribly distracting and disturbing thoughts passing through their mind when they were trying to focus on God. On another site, there was a discussion on how to deal with lustful thoughts as a young Christian. These sites were set up and run by young teenagers. Yet, there was some refreshing maturity in their discussions as they searched for solutions to real problems they were facing as young Christians.

Young Christians on MySpace Display Maturity

On one site, concerning a difficult subject that I will not name here, many of the young people had approached their pastor or pastor's wife with a question about the problem they were facing. However, because the subject was embarrassing, most had received curt answers that didn't help their situation. However, in the blog and comments that followed on the MySpace.com site, the many young people that contributed thoughts on the subject did a great job of being mature and worked together for some practical solutions for each other in this area of struggle. I was thrilled to see that this kind of spiritual maturity is being displayed by our young Christians on MySpace.com.

MySpace.com is a mission field. In my opinion, it is a place that we need to rush into—instead of running away from. Yes, of course there are concerns about security, what our young people are viewing and reading and predators. However, there are also things parents can do to protect their children. For example, you could purchase one of the many programs that are on the market that allow you to monitor what your child is typing (including passwords.) Or, check out this month's "Press Release" column and read about the new product from Covenant Eyes that will provide accountability for MySpace.com, so even if you are browsing or viewing sites that are inappropriate, there is some accountability. Parents must take an active role in monitoring what their children are doing on the Internet, including MySpace.com.

Youth Ministers Could Use MySpace

Imagine how a youth minister could use MySpace.com. Since young people in his church are most likely already using it, he could have a meeting for parents and educate them on the tools they could use to monitor their children and help keep them safe.

He could disciple his young people on how to share their testimony, and then encourage them to post their testimony on how Christ has influenced their lives on their MySpace .com site. He could create a linkable group that once someone began to read about one young person in the group that they could easily read others. MySpace.com has the ability for people to search for others in their community, or EVEN that attend their own high school. Young people looking to connect with other young people will find these Christian youth and that will start relationships and discussion for those who are seeking Christ.

Imagine a youth minister or mature Christian youth that searches for blogs and sites on MySpace.com that are in their community, and post messages within the ongoing discussions that reflect a Christian viewpoint!

Imagine a youth minister that looks through Christian blogs and uses some of the honest questions and discussions there as discussion starters for his own youth meetings. Imagine partnering with young people within his church and set goals to use MySpace.com as a mission field to reach their own community, their city, their state, the USA and even others around the world. Sounds like a "great commission" doesn't it?

A Social Networking Web Site for Muslims Is Growing Rapidly

Anayat Durrani

Anayat Durrani is a Los Angeles journalist who has been widely published worldwide in magazines and newspapers.

Step aside MySpace, Facebook, and Xanga. A new social networking site has joined the ranks. You won't find cursing here. Dating through this site is not permitted. And female members can only post photographs of themselves wearing a headscarf. Welcome to MuslimSpace.com, a new and rapidly growing social networking site catering to the 1.2 billion Muslims worldwide.

MuslimSpace is the brainchild of Mohamed El-Fatatry, a U.A.E.-born Egyptian professional Web developer, designer and programmer living in Finland. A former MySpace user, El-Fatatry created MuslimSpace in March 2006 because he said he was tired of the un-Islamic content of popular social networking sites, such as adult advertisements, and the aggressive behavior of some of their users, not to mention the racist and offensive comments he sometimes encountered on such sites about Islam and Muslims.

"We wanted to make a site where Muslims and non-Muslims from all around the world could come together and enjoy a clean and friendly environment on the Internet from the comfort of their own homes," said El-Fatatry.

The structure of the Web site is similar to other social networking sites. Members can create and personalize their own Web pages, their favorite audio and video, join groups on the site, post pictures and polls, and create blogs.

The personal Web pages on MuslimSpace appear far tamer than those on MySpace. Many members appear in photos in religious and cultural clothing, some with faces covered, exposing eyes only, and their profiles are less detailed. Pages have Islamic ads and Islamic or political polls. They are mainly designed with nationalistic or Islamic content and journal entries seem more Islamically oriented. Verses of the Qur'an are quoted throughout the site on member pages, and audio and video are Islamic in nature.

MuslimSpace Has Strict Rules

"MuslimSpace does not allow any content or images that is even remotely sexual or of adult nature," explained El-Fatatry. "We also don't allow ideologically charged, violent, or controversial opinions, posts, pictures, etc."

Many [MuslimSpace] members appreciate what they deem as a clean, family-friendly atmosphere, including those who are not Muslim.

MuslimSpace is run with a strict set of rules. The site's terms and conditions state that Islamic teachings should be respected at all times. El-Fatatry said that members who engage in political activity "need to be careful not to preach or glorify violent, illegal or criminal acts, otherwise they get banned from the Web site."

El-Fatatry and a group of dedicated student volunteers carefully monitor the site. And they take their work seriously. One member used the word "crap" and was kicked off the site. He was later readmitted.

"At MuslimSpace we always attempt to maintain reasonable standards of decency while remaining flexible enough to allow users to benefit from online social networking," said El-Fatatry. "We have to walk the very thin line between being too conservative or too lean. But we hope the Web site caters to as many Muslims as possible."

The strict rules do not seem to bother any of its 14,000 and growing members. Many of these members appreciate what they deem as a clean, family-friendly atmosphere, including those who are not Muslim. Jennifer, a Christian, who preferred not to use her real name, is a former MySpace.com user. "The main reason [I joined MySpace] was to keep in touch with my immediate friends and after a while I just got a lot of men that I didn't know wanting to get to know me or hang out and I didn't like that."

She said she turned to MuslimSpace after a Muslim friend sent her an invitation to join the Web site. She became a member out of curiosity and also because she wanted to learn more about Islam and about what was going on in the world. "I love that I joined because it's such a safe network and, unlike MySpace, I am actually benefiting from this site," she said. "I don't have to worry about receiving junk, or about stalkers."

Members Come from Around the World

While the majority of its users come from the United States, MuslimSpace draws a sizable number of members from around the world.

Religious-themed social networking sites are the latest to jump on the MySpace bandwagon.

Shahrir Saadun logs onto MuslimSpace from Malaysia. Saadun said he initially joined MuslimSpace to spread the word about a local group he belonged to called, WTV8, which seeks to balance news covered by the media in Malaysia. Though he is now no longer associated with WTV8, he said he continues to use MuslimSpace as a means to connect with Muslims worldwide, particularly to spread news about Muslims, from Muslims to Muslims. "I believe the media is lopsided and biased when it comes to news about Islam and

Muslims," said Saadun. "I intend to spread as much news as I possibly can from the different sites available [through MuslimSpace.com]."

Zaheer, a MuslimSpace.com member living in India, used to be a member of Friendster.com. He joined MuslimSpace because he liked the strong emphasis on his faith and the ability to make friends with Muslims worldwide. "Undoubtedly Islam is the main topic for me. I used to write forums about different topics related to Islam, like on science, medicine in Islam, the meaning of jihad, etc.," said Zaheer.

MuslimSpace has several groups that members can join. Zaheer is a moderator for two of the groups: "Muslims-United" and "Asians." "I created the Muslims-United group to bring closer the Muslims of different sects because somewhere we all are one and I dislike the hostility between sects," explained Zaheer.

Social networking sites are exploding in popularity all over the Internet. MySpace.com reached new heights with 50 million visitors in May, according to comScore Media Metrix. Peter Daboll, president and CEO of comScore Media Metrix said it is unlikely the popularity of such sites will wane in the near future. Social networking is "a phenomenon we're seeing not only in the U.S., but also around the world," he said.

Communicating with like-minded individuals . . . fulfills a basic human need.

Sharing Common Experiences Fills a Need

Religious-themed social networking sites are the latest to jump on the MySpace bandwagon. There is the Christian social-networking site MyPraize.com and Jews have Shmooze.com, among others. But, why are social networking sites so popular?

James H. Fowler, Associate Professor of political science at the University of California, San Diego, an expert on social networks, said it comes down to three words: freedom of expression. "Art fulfills a basic need and the possibility that something you post about yourself can be viewed by billions of other people is quite appealing," said Fowler. "Add to that the ability to form specialized communities, for example with Muslims, where it is possible to share common experiences and you have a winning combination."

Fowler said communicating with like-minded individuals also fulfills a basic human need. Human beings tend toward "homophily," which means a "love of similar things," he said. And while most people are used to hearing that "opposites attract," in fact the opposite is most often the case. "Scientists who study social networks know that we tend to choose to spend time with people of the same political and religious beliefs, the same ethnicity and gender, even the same health characteristics and behaviors," said Fowler.

The growth of MuslimSpace indicates it is filling a need for social networking among Muslims worldwide. El-Fatatry said the site is growing by 100–150 members a day. After the U.S., the bulk of its members are logging in from Malaysia, the U.K., Canada, Turkey, Egpyt, Singapore, Australia, and Germany.

"Members of the MuslimSpace community are very diverse. Similarly, the reasons behind joining the Web site also vary widely," said El-Fatatry. "Meeting new friends from around the world seems to be the most popular reason for joining."

Overall, El-Fatatry said, the feedback to his site has been "outstanding." As with other sites, it has had its share of criticism, as well as hacking attempts, DDoS attacks, and racist harassment aimed at disrupting the Web site. However, with the help of members who use the site's report feature, Muslim-

Space has been able to overcome these issues. Meanwhile, the site continues to draw a steady stream of Muslims who can't seem to get enough.

"More Muslims outside the U.S. are getting into Muslim-Space and most of us have basically the same purpose I think and that is a love for Islam," said Zaheer.

Saleh is an American convert to Islam, who logs on 1–2 times a day. He joined MuslimSpace as a chance to meet Muslims and others from around the world. Saleh exchanges messages with people as far-flung as Japan, Malaysia, Gambia, Jamaica, Canada, Finland and Iran. Most of his communications are with Muslims, but many also are Jews and Christians. His page is a mix of personal, religious and issue-oriented items. "Beyond my own pleasure, I see this site as an opportunity for Muslims to grow in their knowledge of and respect for the variety of understandings and practices which are found in Islam, to make international friends, and for non-Muslims to learn more about Islam," said Saleh.

Seeking to promote an understanding of Islam was another reason El-Fatatry created the site. He said Muslims are often portrayed in a negative light and that Islam is in general a misunderstood faith. Apart from providing Muslims with a platform for social networking, he hopes at the same time his site can help change perceptions and build bridges. "We are hoping that the site provides a medium for tolerant and mature debate between Muslims and non-Muslims, so they can see the true face of Islam," said El-Fatatry. "Non-Muslims who wish to interact with Muslims, know more about Islam, or find out more about different cultures would probably enjoy their stay at MuslimSpace."

CHAPTER4

How Is Online
Social Networking
Changing Society?

Chapter Preface

When online social networking began in the 1980s, some of the pioneers involved remarked on the profound changes it made in the way people communicate. But at that time, it affected very few people; social networking communities were small. Members of one such community did not have online access to the members of another; they could not even send mail to them electronically. Furthermore, users of public commercial systems had to pay by the minute for the time they were online. So, although the potential impact of social networking was great, it did not have any major effect on society as a whole, nor did it significantly change the way most people interacted in their daily lives. Now, for the first time, a large enough percentage of the population is involved for such changes to begin.

Many people welcome new customs and new ways of interacting. They are enthusiastic about the opportunities online social networking offers, and do not feel any regret about the abandonment of what seemed "normal" in the past. Obviously, this is true of young people who have grown up with computer technology and to whom contact online *is* the norm. But it is true of some older people, too.

Others, however, dislike the trends they see developing—not merely the presence of criminals and offensive material on social networking sites, which everyone deplores, but the new and unprecedented attitudes toward traditional values such as privacy. They are dismayed by the fact that many users of these Web sites have no desire to keep details of their personal lives private and are willing to restrict access to their profiles only when convinced that predators may be watching. Furthermore, they feel that if habitual online interaction causes young people not to develop the skills important for face-to-face interaction, this will be a serious loss.

Still other people have neither a positive nor a negative view of social networking's impact on society. They observe the changes that are occurring without passing judgment on them. It is too soon to know how extensive, and how lasting, these changes will be. Probably there will be more changes that are not yet foreseen, just as new media in the past, such as television—and for that matter, the printing press—brought about social and cultural upheavals that were not apparent to their earliest users.

In his 1993 book *The Virtual Community*, published before the World Wide Web was accessible to the public, Howard Rheingold wrote, "My world today is a different world, with different friends and different concerns, from the world I experienced in premodern days. . . . Not only do I inhabit my virtual communities; to the degree that I carry around their conversations in my head and begin to mix it up with them in real life, my virtual communities also inhabit my life." He went on to speculate that "CMC [computer-mediated communications] will be in some way a conduit for and reflector of our cultural codes, our social subconscious, our images of who 'we' might be, just as previous media have been."

This is what is now happening with the advent of huge networking sites such as MySpace, inhabited not by just a few technology enthusiasts, but by millions of average people to whom life online is becoming less and less separate from "real" life. What Rheingold wrote in 1993 is even truer today: "Something big is afoot, and the final shape has not been determined."

The Difference Between "Real" and Online Is No Longer Clear-Cut

Amanda Gefter

Amanda Gefter is a writer for the British magazine New Scientist.

If the Web was once an enormous library, it is now a vast conversation. Transmitting information from one person to another has never been easier. Everyone can participate. Young people now communicate more through social networking websites than through email. Instead of keeping diaries, they keep blogs; instead of photo albums, they have Flickr.

While older adults go online to find information, the younger crowd go online to live. The boundaries between private and public and between offline and online are blurring, and there is a widening generation gap between adolescents growing up with social technology and adults who find it foreign and unsettling. Welcome to the MySpace generation.

A Rapid Change

It all happened remarkably quickly. The first social networking websites were born just three years ago, aimed at providing online forums where friends could connect. A year later online social networking was a fully fledged phenomenon. Today it is the face of the Internet. Social networking websites have evolved from something to visit in your spare time to an integral part of daily life that many today cannot imagine living without.

If you're unconvinced, take a look at the numbers. Friendster, one of the pioneers of online social networking, now has

Amanda Gefter, "This Is Your Space," *New Scientist*, vol. 191, September 16, 2006. Copyright © 2006 *New Scientist*, Reed Business Information, UK, Ltd., a division of Reed Elsevier, Inc. Reproduced by permission.

more than 30 million members. Bebo, launched only last July, has 25 million members and is the number one social networking site in the UK. Then there is the mother of all networking sites, MySpace. Purchased last July by Rupert Murdoch's News Corporation for $580 million, MySpace has just registered its 100 millionth member [159 million by March 2007]. In July [2006] it was ranked number one website among US internet users, receiving more hits in a one-week period than even Google. These figures suggest that online social networking cannot be dismissed as a passing trend. Socialisation, rather than information, has emerged as the primary use of the Internet.

What does it all involve? On a typical networking site like MySpace, you create an individual profile detailing your age, location, whether you are single or in a relationship, gay or straight, plus your general interests, favourite music, movies and books. You might upload photographs of yourself and write daily journal entries. You build up a set of online friends, each of whom will have their own set of friends, and so on. On Friendster, when you view someone's profile the website shows you how you are linked to each other: Bob is a friend of Jane who is a friend of Mark who is your friend. Through varying degrees of separation you are soon connected with hundreds, thousands, even millions of people. Users can communicate with each other in many ways. On MySpace many people post messages on a "bulletin board" that is automatically sent to everyone in their network, or they can post a personalised comment on someone's profile, which is displayed for everyone to read. For one-on-one conversation, they can send direct messages through an internal email system, or chat in real time via instant messaging.

Blogosphere

Another kind of online social networking environment is the so-called blogosphere. Blogs are web pages where individuals regularly post their personal views, like a personal journal or, depending on the blogger, a newspaper column. For teens, the most popular place to blog is a site called LiveJournal, where people keep online diaries for others to read. It has 10.8 million users from all over the world, most of them females around the ages 17 to 19.

The line between what is real and what is virtual is beginning to fade.

Blogging extends well beyond teen diaries, however, according to a website called Technorati, which monitors the blogosphere. It says it is currently tracking 51.3 million blogs worldwide, and claims that 75,000 new blogs are created every day—that's almost one per second. The blogosphere is 100 times bigger than it was three years ago, a doubling in size roughly every six months.

The blogosphere is a good example of how interactive the Internet has become. On most blogs you will find what's called a blogroll—a list of links to other related blogs. Through these links blogs form clusters, or communities, based on shared topics and readership. Conversations develop within and between blogs as readers post comments on what others have written—a phenomenon of huge interest to researchers who study how ideas and information spread among internet users. A new technology called trackback has made the inter-blog network more visible by alerting bloggers every time another blogger creates a link to their site. A blog's importance in the overall network is gauged not in terms of traffic to the site but in the number of inbound and outbound links. For instance, the political blog *The Huffington Post* is ranked number four by Technorati, with 63,918 links from 13,151 other blogs.

Does Online Networking Change Real-Life Social Behavior?

Online social networking, it seems, is penetrating all areas of life and all age groups, even if it is most prevalent among the under-25s. So with people spending so much time communicating online, how is this changing real-life social behaviour? Recent studies have drawn contradictory conclusions. Some claim that individuals who socialise online become more social, extrovert and happy, while others claim that such people draw away from their family and friends, break social ties and become isolated and depressed. So which is it?

The distinction between "real life" and online is no longer clear-cut. The original social websites, such as chatrooms and online games, drew in people who wanted to escape their daily lives, if momentarily, to interact with strangers or to play a fantasy role. They used anonymous screen names and were often represented by avatars or cartoon-like characters. This has changed. Users of social-networking websites are no longer anonymous—they have real names, jobs and relationships. They are represented by photographs. Young people using sites like MySpace switch easily from their real world to their virtual one, and the people they interact with are largely the same.

Despite the media hype, young people on social networking sites are interacting for the most part not with strangers, but with friends from their real life. Thus their online social life doesn't detract from their real one, as the two are simply different manifestations of the same network of friends. Online socialisation is just an extension of the kind of interactions that people have daily by phone, text message and email, so the line between what is real and what is virtual is beginning to fade.

Going Mobile

That line will fade even more as our internet portals go mobile. It is beginning to happen already, and in just a few years the idea of having to sit in front of a computer to access your social network will seem archaic. For instance, a new mobile device called Helio—a phone, camera, video camera, MP3 player and internet browser rolled into one—features "mobile MySpace" that lets you keep in constant contact with your social network.

[In 2005], Google purchased the mobile social network service Dodgeball, which "uses technology to facilitate serendipity." Dodgeball members can send text messages to the site announcing their location—"I'm at the coffee shop on 5th street"—whereupon the service finds any friends in that user's network who happen to be in a 10-block radius and sends them a message suggesting they stop by for an impromptu coffee. Entire cities, including Philadelphia and San Francisco, are transforming themselves into Wi-Fi hotspots, and as wireless devices get better and cheaper, you'll be able to carry your entire social network with you all the time. All of this gives rise to the inevitable existential question: will we be able to have real-life "experiences" if the network isn't watching?

Social Networking for Scientists

Social networking is not just about friends and recreation—it's also starting to affect professional life. Take science. Scientists have used interactive sites such as newsgroups and online bulletin boards since the early 1990s, but the new generation of networking tools is taking that further. Many scientists now use blogs, and the number of science blogs is on the rise—Technorati listed 3329 at the end of August. Science blogs serve a dual purpose. First, they connect scientists to other scientists, serving as modern-day intellectual salons. Formal scientific papers are now even beginning to cite blogs as references. Second, they connect scientists to the general public, offering a behind-the-scenes look at how science progresses.

Meanwhile, other kinds of social networking sites for scientists are emerging. A website called Siphs provides a forum for collaboration and information exchange to around 200 biology and biomedical researchers in 30 countries. If you log on to the site you'll find posts like, "Does anyone know of a good tool that will predict active sites in a protein of interest given an amino acid sequence?" (It seems a world away from MySpace posts like, "Has anyone seen Ashlee Simpson's new nose? Smokin!") Although Siphs is still small, it looks like the beginning of a growing trend: science is fundamentally a collaborative endeavour and online networking is bound to play a major role.

Another technology set to transform scientific research is social bookmarking. Look, for example, at Connotea, created by the publishing group behind the journal Nature, which you can add to the toolbar of your Web browser. When you come across an interesting reference, Connotea saves the link. You then tag your references with keywords, which lets you share your bookmark library or just a few references with colleagues.

The Future of Social Networking

Where will social networking go from here? We can get an idea by looking at some of the latest networking websites, which focus on a specific area of users' lives as a point of common interest. Facebook, launched in early 2004, is specifically for students, and has about 8 million members, two-thirds of whom visit the site every day. LinkedIn is for job networking, and then there's LibraryThing, the "MySpace for bookworms." Here, your profile is your entire book collection, and you can check out who shares the same books and link to their bookshelves.

MySpace connects individuals through friends, Facebook through schools, LinkedIn through professions, and LibraryThing through books. A single individual could be a member of all four sites and many more. It seems inevitable that a meta-network linking together all the various social networking sites will emerge—and an individual's full identity, shown from all sides, will live online. We will carry this meta-network with us in small wireless devices so that our virtual identities become seamlessly integrated with the real world. We will be more autonomous and mobile than ever, and at the same time discover an unprecedented form of collectivism. For the MySpace generation, this won't seem strange at all.

What Used to Be Private Thoughts Are Now Made Public

Tara Bahrampour

Tara Bahrampour is a staff writer for the Washington Post.

Emily Butler used to keep a pen-and-paper diary. But after her mother found it, the Arlington teenager started pouring out her feelings online.

"When there were days when I just needed to rant, it felt good," said Butler, 16, a sophomore at Yorktown High School in Arlington who started a blog on the site Xanga a couple of years ago. "I'd come home after school, and I'd spend, like, an hour typing in everything I did all day."

Butler added: "Once I discovered, like, posting online, it definitely became, 'Why would I write it in a book?'"

Online diaries have become a well-known phenomenon in recent years, with teenagers and young adults attracted to the genre in huge numbers. Raised on the Internet and reality television, these diarists make their writing accessible to friends, acquaintances and, often, to hundreds of millions of World Wide Web users. Many include their full names and school names.

Parents, teachers and police constantly urge young people not to reveal too much about themselves online. They warn that personal disclosures might be read by college admissions officers and potential employers, not to mention stalkers and pedophiles. The risks were underscored in a highly publicized 2005 Virginia murder case in which investigators looked for clues in the online journals of college student Taylor Behl and her killer.

Tara Bahrampour, "On the Web, 'Dear Diary' Becomes 'Dear World'," *Washington Post*, January 2, 2007, p. B02. Copyright © 2007 *The Washington Post*. Reprinted with permission.

Online Diaries Are Popular

But a review of major blogging and social-networking Web sites shows that online diaries remain popular for teenagers, and interviews with experts and young diarists such as Butler help explain the psychology behind going public with what used to be private thoughts.

Young diarists say the [online] journals connect them to a broader community, help them navigate the complexities of friendship and romance and allow them to vent.

A few examples from area high school students:

"Unfortunately I feel very distant from everyone. . . . Maybe it's just how I function. I think its probably my worst flaw."

"i feel she could be the one i know it is crazy because well i am 18 and all that but i really do i am just scared i have never let someone get as close to me as i have let her."

"i feel . . . invisible."

A Manassas-area teenager writes of her sadness and loneliness after seeing her father choke to death on a piece of steak. A Montgomery County high-schooler recounts the bliss of falling in love for the first time and then, months later, the anguish of breaking up. A Prince William County girl sent to a group home laments that old friends seem more distant.

Of course, it is hard to know how many of these diary entries represent truth as the writers see it, fantasy or something in between. Regardless, young diarists say the journals connect them to a broader community, help them navigate the complexities of friendship and romance and allow them to vent.

Teenagers also use online diaries to spread information quickly. "You can get to a lot of people all at once," said Colton O'Connor, 19, a recent graduate of the Thomas Jefferson High School for Science and Technology in Fairfax County who is now a freshman at the College of William and Mary. "Like, a phone call only gets to one person at a time."

It's impossible to determine how many young people keep online diaries, but companies that operate major blogging and networking sites—such as Xanga, LiveJournal and MySpace— say the numbers of teenagers and young adults that use them are in the millions.

Private Thoughts in Public Forums

Young people point out that posting private thoughts in a public forum has become more acceptable with the rise of cultural phenomena such as PostSecret, a popular Web site that displays postcards emblazoned with senders' secrets. Xanga, LiveJournal and MySpace all give users the option of making their blogs accessible only to approved readers. Some also keep "secret blogs" on which they enter intimate thoughts that, in the old-school tradition, are meant for the writer's eyes only. But many young people prefer to lay it all out for the world to read.

Blogs let writers interact while avoiding the emotional risks of one-on-one conversation.

Rochelle Gurstein, author of *The Repeal of Reticence*, a book about the erosion of privacy in the United States, said the blogs seem to reflect an "unprecedented change" in teenagers' sense of modesty.

"Not long ago, young people would die at the prospect of their mother or their friend discovering" their diaries, she said. "The teenage girl that used to be the most vulnerable, protected member of society is now unsupervised, left to her own devices, with access to the Internet, and what does she do? Broadcasts to the whole world to see her in her most vulnerable moments."

But O'Connor, who has kept a LiveJournal diary for more than two years, said blogs actually protect vulnerabilities by allowing for a more polished presentation of self. "You can take three minutes to lay your thoughts out and think about it before you send them," he said.

His older brother noted that blogs let writers interact while avoiding the emotional risks of one-on-one conversation.

"This generation is worse at talking face to face," said Jeremy O'Connor, 23, a recent Virginia Tech graduate who has kept a LiveJournal diary for five years. "Everything everyone's writing online, they want it there because they want it to be read by someone. . . . Having someone read your secret feels better."

He added that blogs allow people to communicate obliquely, "writing, 'I like so and so,' and knowing it's going to get back to that person without having to talk to that person."

Many young bloggers say they don't think people other than friends are reading their journals. Some contend that the Internet is a safer place for their inner thoughts than a book that can be found by parents or siblings.

Parents are less sure.

The O'Connors' mother, Karen, said she was appalled when her four children started keeping online journals. "I just thought it was terrible, horrible. I just couldn't imagine why you would put your feelings and personal comments on something that just went out there."

A Social Buffer

She now sees good and bad in it. "You probably know your friends better because they put everything on LiveJournal. But you're missing all the excitement and fun" of face-to-face interactions.

Gerald Goodman, professor emeritus of psychology at the University of California at Los Angeles, said young bloggers are following a deep human impulse. "This is practically genetic, this need to be known by another human," he said. But Goodman said he worries there is a downside for those who rely too much on such communication.

"It's not real—it's like phone sex or something; it's partial," he said. "As they grow up, what happens to how they manage their vulnerability and their disclosure and their risk-taking in human relations? Is this going to do something that we can't predict yet about the way they're willing to take risks to get close?"

Butler acknowledged that relating online provides a social buffer. But that is part of the appeal. "Saying, like, 'Hey, do you like me?'" she said. "In person it would be the most awkward thing in the world."

Breaking up online is also all right, she said. Then she reconsidered: "Breaking up online is so sixth grade. Like, by eighth grade you should at least call them."

Online Social Networking Has Altered the Rules of Social Interaction

Dan Bobkoff

At the time this article was written Dan Bobkoff was an executive editor of the Wesleyan Argus, *a paper published by the undergraduates of Wesleyan University.*

When Susannah Fox '92 attended Wesleyan, she often spent her Wednesday nights at "Blues and Brews," a weekly house party held by Eric Halperin '92 and his friends. On one of these Wednesdays—Robert Cray playing in the background—Fox and Halperin met by the keg and hit it off. The trouble was, after that night, Eric disappeared.

"I would scour the library and would have to walk all around campus, looking, trying to find him," Fox said. Lucky for them, they did find each other, and eventually married in 1998. But, she wonders if it would have been easier to get to know him if they were in college today. "I would never have called and neither would he for a 'date,'" she said. "If I had just gotten his screen name or email, our impression is that it would have been so much easier to connect."

That's probably true. Today's students can contact each other with cellphones, e-mail, text messaging, Instant Messenger and the most recent addition, facebook.com (known to most as "Facebook"). These only became widely available in the last decade.

Facebook's adoption among students has been astounding, with over 2.2 million users at 496 colleges and growing. There are 2,985 users at Wesleyan, including students, alumni, faculty, and staff. Anyone with a valid Wesleyan email address can create an account and profile on the site. Given that level of acceptance, it suggests that students are comfortable forfeiting their personal information for the social gains the site affords. But are they putting themselves at risk?

Students walk around campus knowing information about strangers, acquaintances, and crushes without ever having to interact with them.

For the uninitiated, Facebook is a social networking website where students post profiles with photographs of themselves, and a wide range of information from cellphone numbers, addresses, Instant Messenger screen names, to sexual orientation, whether they're in a relationship or not (and if so, with whom), to favorite movies, music, quotes, and a free response section. Some students will also post links to online photo albums and diaries that might contain pictures of underage drinking and the students' innermost thoughts. The site also displays a list of a user's friends and lets those friends leave public comments on your profile.

The site, then, is helping to change the ways students forge friendships and relationships, and is able to do so because of students' willingness to post copious amounts of personal information for anyone in the school community to see. Now, students walk around campus knowing information about strangers, acquaintances, and crushes without ever having to interact with them.

And that information can be addictive. "I once tried to go a day without signing onto Facebook and failed miserably," said self-described Facebook addict Elissa Gross '08. "I went until 8 o'clock at night."

The site is fun, useful, and a great procrastination tool, but some see it as part of a shift in how we interact with one another. "We're coming from a tradition going back tens of thousands of years in which when people met, they didn't know anything about each other, or just [knew] what a friend told them," said Paul Levinson, Professor and Chair of Media Studies at Fordham University, and the author of *Digital McLuhan: A Guide to the Information Millennium.* "What I think something like the Facebook does is that it gives everyone a little dossier, and it does profoundly change the rules of engagement when they do meet," he said. "It obsolesces small talk."

Students Believe They Are Safe

Most students I spoke with did not believe they were sacrificing safety by being on the site. "I figure that with the Internet, you can find any information you're looking for," said Ruby Ross '08, whose Facebook profile lists Flamenco dance and chocolate soymilk as interests.

Most people don't understand the implications of their actions on the Internet.

Nor is Allison Wilcox '07 particularly worried about people using her information maliciously. "It's a small enough campus that if you want to stalk someone, you don't need Facebook to do it," she said.

Students' laissez-faire attitude toward privacy on the site troubles Parry Aftab, an internet privacy and security lawyer and executive director of WiredSafety.org. She said students feel too secure on Facebook. When presenting themselves online, students are inclined to assume that only savory students will be checking their information and contacting them. "There's a disconnect between the construction of who's reading your post and the reality," she said. "The good people are not necessarily who are reading it."

Aftab, who uses teen and collegiate volunteers to help stay abreast of net trends, says her organization often receives word of abuses from Facebook users. "We get reports all the time of cyberstalking and harassment," she said. "A lot of them become sexual very quickly." Aftab also cites instances of identity theft and cyberstalking among Facebook users.

However, Chris Hughes, a spokesman for facebook.com, said that the site rarely receives reports of stalking or other harassment. "Most complaints about 'stalking' simply require a readjustment of a user's privacy settings and a warning message sent to the user accused of inappropriate behavior," he said. "We think our users are pretty savvy about what information they give out to their peers."

No one I spoke with knew of any illicit behavior through Facebook at Wesleyan.

Employers [are] perusing the profiles of potential employees, comparing their online self-portrayals to their resumes.

Fox, who is now associate director of the Pew Internet & American Life Project, a nonprofit organization that studies how the internet affects areas of our lives, says that most people don't understand the implications of their actions on the Internet. "They love the convenience [of the internet]," she said. "For the most part, nothing bad has happened to them so they continue to do these high-trust activities."

Her colleague, Steve Jones, a senior research fellow at Pew Internet, and a professor of communication at the University of Illinois at Chicago, said that many of his students didn't consider the consequences of their online profiles until after they put them online. "Facebook seems safer because it's restricted to your own campus," he said.

Jones warns, however, that there could always be unexpected consequences. "Let's say you put something in your profile about how you're a real party animal, and 20 years later, you're running for political office and someone discovers it," he said. "On the other hand, I don't think everything everyone puts up there is true."

True or not, there are other emerging consequences for posting a risqué Facebook profile.

University of Wisconsin-Madison freshman Jill Klosterman discovered that campus employers were perusing the profiles of potential employees, comparing their online self-portrayals to their resumes. "I was told by a [Wisconsin Union] employee that ensuring that my Facebook profile was employer-safe was 'strongly advised'," she said.

Jones likens employers using Facebook to basing hiring decisions on hearsay. It seems clear, though, the days of students seeing the site as purely their domain may be numbered.

Professors Are Joining Facebook

At Yale, for instance, professors have begun creating their own profiles, often at the insistence of students who are enamored with a particular professor. The reaction to professors joining Facebook there seem to be mostly warm, according to *Yale Daily News* writer and freshman Josh Duboff, who wrote an article about the phenomenon. "Most people seem to think that there's nothing on their Facebook profiles that is so private that they are worried about faculty members seeing it," Duboff said. "The professors seem to have an amusing attitude toward the whole thing."

A great many of us do not fear incursions on our privacy, but rather want to be found.

About 30 Yale professors have joined. At Wesleyan, a quick search didn't turn up any active faculty, but that could quickly change. At the University of Iowa, even the school's president is on the site now. Facebook may shift to become an online reflection of the physical school community.

Levinson, who plans to join the site at Fordham soon, is unsure how this evolving community will take shape. "It could bring faculty and students together, but it could also undermine some of the advantages and camaraderie students now have with it," he said. He points to sexual preferences as one such trait students might not want professors to view. Facebook does offer privacy preferences that allow students to tailor who sees aspects of their profiles, however.

"I think most of our users are intelligent enough to know that any information they're posting to a site for their college will, in fact, be seen by other members of their collegiate community," Hughes said.

Aftab, who actually likes Facebook, said students should just be prudent about what they post. "You shouldn't use it as a diary," she said. Aftab warns that students should also avoid being sexually provocative, or taking strong political or other opinions "that are going to evoke a strong reaction from other people. Otherwise you're putting a target on your back," she said.

Risky or not, Facebook's popularity is a direct result of students' willingness to forgo privacy to belong to a community. Perhaps the most profound result of the rise of Facebook is that it demonstrates a great many of us do not fear incursions on our privacy, but rather want to be found. Many, overtly or not, hope people are searching for their profiles and finding them interesting. Otherwise, why belong to the site at all?

Or, as Alan Yaspan '08 responded when asked if he'd ever been stalked online, "I could use an ego boost."

"I think pretty much everyone that posts all this information wants people to contact them," said Dan Stillman '04. "They just don't want to seem too eager about it." Stillman is the co-creator of WesMatch, a competitor to facebook.com that allows users to find students who are compatible with them by filling out a questionnaire and then checking a list of whose answers most closely resemble their own.

People Now Research Each Other

The etiquette of these technologies is still developing. Facebook, WesMatch, and the Internet as a whole, have legitimized researching our acquaintances, classmates, and crushes.

In the information age, we expect to do the same kind of research on a potential girlfriend or boyfriend as we would [on] a new toaster.

"Thanks largely to Google and the Web in general, there's an expectation today that, if you want to find out information about a person, you'll be able to just load up a web browser and type in a name," Stillman said. This expectation is recent, however. When he was a freshman, people were surprised if he knew about their backgrounds before they spoke. "Today, I think many people are surprised if someone hasn't looked them up," he said.

Meeting someone they had researched on the web, some students find themselves asking questions they already know the answer to. It's not quite socially acceptable to admit you've been researching the person. There is a disconnect between the acceptability of online snooping and the permissibility of using that information. Like many students I spoke with, Ross said this has happened to her when meeting friends of friends she had looked up. "I had to pretend my ignorance," she said.

Matt Gregory '07 found himself unexpectedly unnerved when he walked into a friend's room and found her looking at his profile. "She was doing it to get contact info, but it still made me feel uncomfortable," he said. "I felt that was something she could have gotten in a more personal manner."

Indeed, finding friends has become akin to shopping online. You can see customer reviews (what someone's friends have added to a profile), and even add or cancel a relationship with the push of a button. In the information age, we expect to do the same kind of research on a potential girlfriend or boyfriend as we would a new toaster.

Levinson thinks the site is helping students find each other more efficiently. "The fact is, on the one hand humans have a lot of basic similarities, but when you get beyond that, there is a myriad of very disparate interests," he said. "By making this information available, it encourages people to meet people who are interested in the same things you are."

Gross said the site makes her and others gutsier. "I think college students like Facebook because, like alcohol, it really lowers your inhibitions and timidity in terms of posting or messaging things you would never say in real life," she said.

Gross said that, among her friends, it's impressive to get with someone who chose not to be on the site. "It's like a challenge now, trying to find someone who's so mysterious and elusive that they would avoid this newfound cult," she said.

Anonymity is not just hard to attain on campus now, it is rarely desired, either. Those students who choose to avoid Facebook, won't sign on Instant Messenger, and don't have a cellphone, are seen as modern day Luddites, quietly disrupting the new social order. Many students expect others to be reachable and researchable anytime of day from anywhere. And, with that, come new risks and rewards.

Privacy, already scarce on a campus as small as this one, is almost extinct, but that's just how students like it. As Yaspan puts it, "Nowadays, it seems like you have to have some presence in cyberspace to be anybody at all."

Online Social Networking Is Changing the Way People Mourn

Warren St. John

Warren St. John is a reporter for the New York Times.

Like many other 23-year-olds, Deborah Lee Walker loved the beach, discovering bands, making new friends and keeping up with old ones, often through the social networking site MySpace.com, where she listed her heroes as "my family, and anyone serving in the military—thank you!"

So only hours after she died in an automobile accident near Valdosta, Ga., early on the morning of Feb. 27, her father, John Walker, logged onto her MySpace page with the intention of alerting her many friends to the news. To his surprise, there were already 20 to 30 comments on the page lamenting his daughter's death. Eight weeks later, the comments are still coming.

"Hey Lee! It's been a LONG time," a friend named Stacey wrote recently. "I know that you will be able to read this from Heaven, where I'm sure you are in charge of the parties. Please rest in peace and know that it will never be the same here without you!"

Just as the Web has changed long-established rituals of romance and socializing, personal Web pages on social networking sites that include MySpace, Xanga.com and Facebook.com are altering the rituals of mourning. Such sites have enrolled millions of users in recent years, especially the young, who use them to expand their personal connections and to tell the wider world about their lives.

Inevitably, some of these young people have died—prematurely, in accidents, suicides, murders and from medical problems—and as a result, many of their personal Web pages have suddenly changed from lighthearted daily diaries about bands or last night's parties into online shrines where grief is shared in real time.

The pages offer often wrenching views of young lives interrupted, and in the process have created a dilemma for bereaved parents, who find themselves torn between the comfort derived from having access to their children's private lives and staying in contact with their friends, and the unease of grieving in a public forum witnessed by anyone, including the ill-intentioned. "The upside is definitely that we still have some connection with her and her friends," said Bob Shorkey, a graphic artist in North Carolina whose 24-year-old stepdaughter, Katie Knudson, was killed on Feb. 23 in a drive-by shooting in Fort Myers, Fla. "But because it's public, your life is opened up to everyone out there, and that's definitely the downside."

Her really close friends go on there every day. It means a lot to know people aren't forgetting about her.

It's impossible to know how many people with pages on social networking sites have died; 74 million people have registered with MySpace alone, according to the company, which said it does not delete pages for inactivity. But a glib and sometimes macabre site called MyDeathSpace.com has documented at least 116 people with profiles on MySpace who have died. There are additions to the list nearly every day.

Last Thursday, for example, a 17-year-old from Vancouver, Wash., named Anna Svidersky was stabbed to death while working at a McDonald's there. As word of the crime spread among her extended network of friends on MySpace, her page was filled with posts from distraught friends and affected strangers. A separate page set up by Ms. Svidersky's friends after her death received about 1,200 comments in its first three days. "Anna, you were a great girl and someone very special," one person wrote. "I enjoyed having you at our shows and running into you at the mall. You will be missed greatly . . . rest in peace."

Tom Anderson, the president of MySpace, said in an e-mail message that out of concern for privacy, the company did not allow people to assume control of the MySpace accounts of users after their deaths. "MySpace handles each incident on a case-by-case basis when notified, and will work with families to respect their wishes," Mr. Anderson wrote, adding that at the request of survivors the company would take down pages of deceased users.

Friends of MySpace users who have died said they had been comforted by the messages left by others and by the belief or hope that their dead friends might somehow be reading from another realm. And indeed many of the posts are written as though the recipient were still alive. "I still believe that even though she's not the one on her MySpace page, that's a way I can reach out to her," said Jenna Finke, 23, a close friend of Ms. Walker, the young woman who died in Georgia. "Her really close friends go on there every day. It means a lot to know people aren't forgetting about her."

More formal online obituary services have been available for a number of years. An Illinois company called Legacy.com has deals with many newspapers, including The New York Times, to create online guest books for obituaries the papers publish on the Web, and offers multimedia memorials called Moving Tributes for $29. But Web pages on social networking sites are more personal, the online equivalent of someone's room, and maintaining them has its complications. Some are frustratingly mundane.

Amanda Presswood, whose 23-year-old friend Michael Olsen was killed in a fire in Galesburg, Ill., on Jan. 23, said none of his friends or family members knew or could guess the password to his MySpace account, which he signed onto the day before he died. That made it impossible to accept some new messages. "There's a lot of pictures on there that people haven't seen," Ms. Presswood said. "His parents have been coming to me for help because they know I know about the Internet. They even asked if I could hack it so I could keep the page going."

The Walkers correctly guessed the password to their daughter's page, and used it to alert her friends to details of her memorial service. They also used it to access photographs and stories about their daughter they had missed out on. "It's a little weird to say as a parent, but the site has been a source for us to get to know her better," Mr. Walker said. "We didn't understand the breadth and scope of the network she had built as an individual, and we got to see that through My-Space. It helped us to understand the impact she's had on other people."

At the same time, Ms. Walker's mother, Julie, wrote in an e-mail message, the family was overwhelmed by unsolicited e-mail messages from strangers offering platitudes and seeking to advise them on how to handle their grief. The family found such offerings unwelcome, however well intentioned. "The grief of our own friends and family is almost more than we can bear on top of our own, and we don't need anyone else's on our shoulders," Mrs. Walker wrote.

Mr. Shorkey said he and his wife remained in touch with their daughter's friends through MySpace. And they visit her Web page daily. "Some days it makes me feel she's still there," he said. "And some days it reminds me I can never have that contact again."

MySpace Is Becoming a Marketing Madhouse

Wade Roush

Wade Roush is a consulting editor for Technology Review.

Web users have created more than 116 million profiles on MySpace, the social-networking site owned since 2005 by Rupert Murdoch's News Corp. As I will explain in a moment, many of these profiles are fake. Still, 116 million is more than the number of people in Mexico and the number of cable TV subscribers in the United States.

Parents and members of the U.S. Congress have begun to take note—and they don't like what they see. Conservative groups fomented a media panic this year over the supposed rash of sexual predators on MySpace and pushed a bill through the House of Representatives—the Delete Online Predators Act (DOPA)—that would cut off minors' ability to access this and other social-networking sites from federally funded facilities like schools and libraries.

In the opinion of experts such as Henry Jenkins, a professor of literature and director of the Comparative Media Studies Program at MIT, the threat of sexual solicitation on MySpace is not as great as many fear. The company has indeed been hit with a high-profile lawsuit over an incident in which an adult molester allegedly met his underage victim on the site. But teens who use the Internet have said in surveys that online "solicitations" often come from people under 25—and are simply ignored. Furthermore, MySpace is likely to get safer: an October Wired News report that as many as 744 registered sex offenders have MySpace profiles will likely push the company to cull such members.

But while MySpace's bad rap as a haven for sexual predators is probably undeserved, there's good reason to be disturbed by the site: it is devolving from a friends' network into a marketing madhouse.

[MySpace] has given members the technological tools to "express themselves" by turning their own profiles into multimedia billboards for bands, movies, celebrities, and products.

Beyond Traditional Advertising

If any social-networking company has found a way to rake in cash, it is MySpace; for example, Google recently agreed to pay $900 million for the exclusive right to provide Web searching and keyword-based text ads on the site. Of course, targeted advertisements distributed by Google and other companies provide the revenue that keeps many Web-based businesses afloat. But MySpace's venture into consumer marketing has gone far beyond traditional advertising. The site has given members the technological tools to "express themselves" by turning their own profiles into multimedia billboards for bands, movies, celebrities, and products. Think MTV plus user photos, bulletin boards, and instant messaging.

I realize that in criticizing a pop-culture mecca frequented by millions of people, I risk sounding just as out of touch as DOPA's supporters. But after spending the last few years chronicling the emergence of social networking and other forms of social computing for this magazine, I had higher hopes for the technology. To me, the popularity of MySpace and other social-networking sites signals a demand for new more democratic ways to communicate—a demand that's likely to remake business, politics, and the arts as today's young Web users enter the adult world and bring their new communications preferences with them. The problem is that MySpace's choice of business strategy threatens to divert this populist energy and trap its users in the old, familiar world of big-media commercialism.

My biggest worry about MySpace is that it is undermining the "social" in social networking. The general expectation when one joins a social network is that its other members are actual people. On MySpace, this isn't always so. The movie *Jackass Number Two* has a profile on the site, as do Pepsi, NASCAR, and Veronica Mars, the CW network's teen detective. The company interprets the idea of a "profile" so broadly that real people end up on the same footing as products, movies, promotional campaigns, and fictional characters—not exactly the conditions for a new flowering of authentic personal expression.

As a site organized around an enormous collection of profiles, MySpace was modeled on Friendster and other earlier online social networks. Users are given pages where they can post self-descriptions, photos, short videos, blog entries, and the like. Every profile includes a list of the other members its creator has "friended," and a comment section where those friends leave feedback. (Most comments are encouraging, casual, and shallow: "Love the new look! How are you not married yet?")

Friendly to Independent Artists

But one feature that makes MySpace different from earlier sites, and evidently more appealing to users, is its friendliness toward independent artists. Cofounder Tom Anderson, who has a background in the Los Angeles arts scene, has said that he and business partner Chris DeWolfe started the site in 2003 because the older social networks didn't give musicians, photographers, digital filmmakers, and other artists adequate ways to promote themselves and their work. From its beginning, then, MySpace has functioned as a public stage. It lets bands and solo musicians create profiles, publicize upcoming shows, and upload their songs, which other members can then embed in their men profiles. Filmmakers can upload video clips. Indeed, the site has become one of the main places where unknown artists go to be discovered by major studios, or at least to develop a base of fans who'll attend shows and buy CDs and DVDs.

In the early days at Friendster, only real individuals could create profiles. Bands were lumped in with other "fakesters," the term coined by Friendster users for profiles created by impostors or dedicated to someone other than the author, such as a pet or a celebrity. The company eventually relented, and fakester profiles became an accepted part of Friendster's culture, often taking on the function of fan clubs.

MySpace, however, has been hospitable to fakesters from the beginning—so much so that it's now perfectly kosher for a company (or one of its fans) to create a profile for a fast-food chain, a brand of soda, or an electronics product. Other MySpace members can friend these profiles just as if they represented people. As of early October, Burger King had more than 134,500 friends, and the Helio cell phone had 130,000.

The large supply of fake "friends" . . . encourages members to define themselves and their relationships almost solely in terms of media and consumption.

Members Declare Cultural Affinities

The fakester phenomenon gives network members a way to declare their cultural affinities. These declarations are a huge part of a member's online identity, according to social-media researcher danah boyd, who is studying MySpace and other social-networking sites for her doctoral thesis at the University of California, Berkeley's School of Information. "It is important to be connected to all of your friends, your idols and the people you respect," boyd writes. "Of course, a link does not necessarily mean a relationship. . . . The goal is to look cool and receive peer validation."

But profiles are about more than looking cool, in boyd's view. She argues that social-networking sites are among the last unregimented environments for young people, places where they're free to explore issues of personal and group identity. Members of such sites "write themselves into being" through their profiles, boyd says, trying out personalities and slowly coming to understand who they are and how they fit in.

Ideally, every networking site would be this liberating. Alas, MySpace tends to herd its users into niches created for them by the mass market. If MySpace members are writing themselves into being through the profiles they friend and the products they endorse, then today's 14-to-24-year-olds are growing up into a generation of Whopper-eating, iPod-absorbed, Hollywood-obsessed Red Bull addicts.

Take BillyJ (not his real handle), an 18-year-old high-school graduate and UPS employee in Louisville, KY. BillyJ smokes Kools, prefers Coke to Pepsi, counts *X-Men: The Last Stand* among his 395 friends, admires New Jersey Nets guard Jason Kidd, likes to work on car audio systems, doesn't have a girlfriend yet, and apparently covets a Ducati motorcycle (his profile features customized Ducati backgrounds, color schemes, and ads). BillyJ may have deeper, more personal interests, but you won't find them on his MySpace profile. It s unclear what he contributes to the network—but as a single 18-to-24-year-old male with his own income and lots of friends, he is a viral marketer's dream vector.

A Platform for Product Placement

In fact, MySpace can be viewed as one huge platform for "personal product placement"—one different from big-media-style product placement only in that MySpace members aren't paid for their services. There's nothing new, of course, about word-of-mouth marketing. What's sad about MySpace, though, is that the large supply of fake "friends," together with the cornucopia of ready-made songs, videos, and other marketing materials that can be directly embedded in profiles, encourages members to define themselves and their relationships almost solely in terms of media and consumption.

This can't be all that social computing has to offer. Older Web-based social networks were launched with serious (or at least creative) missions: LinkedIn is about making business connections, Flickr and Fotolog are for sharing photographs, Meetup is for planning book clubs and campaign events. Of course, there's no requirement that a social network have high ideals. Like television and every other technology that started out as a shiny showroom prototype, social networking will inevitably accumulate some dings and scratches on the road to mass adoption. But if MySpace is to be the face of online social networking, it's fair to ask whether it's making our culture richer or poorer. To date, the only people who are profiting are Rupert Murdoch and his stockholders.

Users of Social Networking Web Sites Are Not Passive Consumers

Howard Rheingold

Howard Rheingold is a writer, speaker and Web developer who has been involved in online communities since the 1980s. He has written many books about the impact of technology on society.

Every year, [author and literary agent] John Brockman asks a question of a widespread community of thinkers and publishes it on Edge.org. I reproduce here my answer to this year's question—"What are you optimistic about?"—in its entirety:

The tools for cultural production and distribution are in the pockets of 14-year-olds. This does not guarantee that they will do the hard work of democratic self-governance: the tools that enable the free circulation of information and communication of opinion are necessary but not sufficient for the formation of public opinion. Ask yourself this question: Which kind of population seems more likely to become actively engaged in civic affairs—a population of passive consumers, sitting slackjawed in their darkened rooms, soaking in mass-manufactured culture that is broadcast by a few to an audience of many, or a world of creators who might be misinformed or ill-intentioned, but in any case are actively engaged in producing as well as consuming cultural products? Recent polls indicate that a majority of today's youth—the "digital natives" for whom laptops and wireless Internet connections are part of the environment, like electricity and running water—have created as well as consumed online content. I think this bodes well for the possibility that they will take the repair of the

Howard Rheingold, "The Tools of Cultural Production Are in the Hands of Teens," *SmartMobs*, January 1, 2007. www.smartmobs.com. Reproduced by permission of the author.

world into their own hands, instead of turning away from civic issues, or turning to nihilistic destruction.

The eager adoption of Web publishing, digital video production and online video distribution, social networking services, instant messaging, multiplayer role-playing games, online communities, virtual worlds, and other Internet-based media by millions of young people around the world demonstrates the strength of their desire—unprompted by adults—to learn digital production and communication skills. Whatever else might be said of teenage bloggers, dorm-room video producers, or the millions who maintain pages on social network services like MySpace and Facebook, it cannot be said that they are passive media consumers. They seek, adopt, appropriate, and invent ways to participate in cultural production. While moral panics concentrate the attention of oldsters on lurid fantasies of sexual predation, young people are creating and mobilizing politically active publics online when circumstances arouse them to action. 25,000 Los Angeles high school students used MySpace to organize a walk-out from classes to join street demonstrations protesting proposed immigration legislation. Other young people have learned how to use the sophisticated graphic rendering engines of video games as tools for creating their own narratives; in France, disaffected youth, the ones whose riots are televised around the world, but whose voices are rarely heard, used this emerging "machinima" medium to create their own version of the events that triggered their anger ([the movie] *The French Democracy*). Not every popular YouTube video is a teenage girl in her room (or a bogus teenage girl in her room); increasingly, do-it-yourself video has been used to capture and broadcast police misconduct or express political opinions. Many of the activists who use Indymedia—ad-hoc alternative media organized around political demonstrations—are young.

My optimism about the potential of the generation of digital natives is neither technological determinism nor naive utopianism. Many-to-many communication enables but does not compel or guarantee widespread civic engagement by populations who never before had a chance to express their public voices. And while the grimmest lesson of the twentieth century is to mistrust absolutist utopians, I perceive the problem to be in the absolutism more than the utopia. Those who argued for the abolition of the age-old practice of human slavery were utopians.

Online Social Networking Enables Users to Be Part of a Global Community

Eston Bond, interviewed by Kevin Farnham

At the time of this interview, Eston Bond was a student at the University of Michigan and was also working as a Web designer, online journalist and technical writer.

[Eston Bond:] I'm still wondering what the end effect of social networking will have on the future of this generation. I think that social networks give us a core perspective on what we're seeing reflect in the rest of the social Web, and it carries with it the centre of the whole Web 2.0 revolution. I read somewhere the other day that the previous communications revolution was when multinational corporations tied the world together and allowed international populations to interact with one another on a mainstream, capital-based level. This was our previous communications era, where people primarily interacted with those around them locally and regionally, but came into contact with those internationally via corporate interaction (or if they could afford the costs of international travel personally). You used to make "pen pals" as social contacts in other countries. Friends in other domestic regions were ones you called long distance every month or so. Sure, interaction existed, but it was sporadic. Costs of maintaining international interaction day-to-day was more for business than something social. With the explosion of relatively cheap computer hardware and even cheaper broadband Internet access, that control of interaction shifted from corporation to consumer, and the marginal cost of maintaining international contacts plummeted. It costs me the same to col-

Eston Bond, interviewed by Kevin Farnham, "Eston Bond on the Future of Social Networking," O'Reilly diyINcite MySpace blog, October 10, 2006. http://myspace.com/oreillymedia. Reproduced by permission of the authors.

laborate with developers in Gothenburg, Sweden as it does to collaborate with developers here in Ann Arbor. The cost is as negligible to me as an individual as it is to an enterprise.

We're seeing that happen socially, too. We can maintain friendships with those in St. Petersburg as easily as those in St. Paul. Social networks also give us seriously increased efficiency in finding those with similar interests, especially so if we have very specialised ones. As more of the world comes online, we also have the capability to run across those with entirely different life experiences yet maybe one very minute common interest, and that microscopic particle of commonality can connect people that would have never been able to speak to each other before, either by social segmentation or geographic separation. A lot of people say that this ability to find niche communities is specialising and isolating us further, and social networks are merely facilitating the increase in comfort of those that find solace in that specific niche. They say that the social network just reinforces closed-mindedness and niche behaviour. I don't believe that this is the case: that social network profile is accessible to everyone by default. Sharing the smallest, weirdest interest with someone can start an intercontinental conversation with someone who has origins in a different culture, who speaks a different primary language.

Social networking is bringing us all closer to a greater humanity by allowing this type of interaction. It's the beginning to the destruction of geographical and political barriers that segregate us and pushes each of us toward being both absolutely, entirely unique yet more knowledgeable of other cultures, making each of us an assimilate of a global culture as opposed to a local, regional, or national one. It begins to tie us together into a greater human cause than those afforded to us by nationalism. It has the power to give everyone a unique voice and a way to find others. Social networking can help us in all of our most human desires and ideals: we can more easily find those to collaborate with intellectually. A social network can help being lonely people together. It can save the lives of the depressed and offer support to the mourning. Our community's size increases drastically, and with it the magnitude of human emotion is amplified.

Social media will do more than affect the teenagers and twenty-somethings. Social media will be the core of the common human cause.

When I tell people that I don't work with social networks in the way that they (common users) do, this is the reason why. I'm not worried so much about using MySpace, Facebook, or any other such network to meet friends and girlfriends; I'm using social networks so *everyone else* can meet friends and significant others *globally*, so they, too, can contribute to a greater cause. I'm extremely fortunate that with my usage of the network in this humanitarian way has the positive impact of allowing me to meet all sorts of people in the way a common user would, and from there I get the extra benefit of being able to experience the type of global society that I'm trying to expand. It's a recursive cause. Like I said earlier: you get what you give.

While my Hack Pack [software for developing MySpace profiles] was superficial, I'm hoping it'll be a springboard into bringing others closer to the global sphere. If I can affect one person into seeing the complexity and beauty of the social network's architecture through the vector of design, they'll affect someone else, and the chain grows exponentially. I can't build a greater social network single-handedly. I can't make it more aesthetically beautiful or more efficient across international boundaries. Everyone working together, though, can build a greater social network, either by simply participating or developing the infrastructure, and that's what I intend to facilitate. If things continue on their current runaway path, we'll see a drastic change in the power structures of society.

Social media will do more than affect the teenagers and twenty-somethings. Social media will be the core of the common human cause and affect anyone and everyone, both individually as well as a population. I can't tell you exactly what the future of social networks will be; I can just tell you what we are capable of making them to be.

Organizations to Contact

The editors have compiled the following list of organizations concerned with the issues debated in this book. The descriptions are derived from materials provided by the organizations. All have publications or information available for interested readers. The list was compiled on the date of publication of the present volume; the information provided here may change. Be aware that many organizations take several weeks or longer to respond to inquiries, so allow as much time as possible.

American Library Association
50 East Huron Street, Chicago, IL   60611
(800) 545-2433
Web site: www.ala.org

The ALA is a professional organization for libraries and librarians. It is strongly opposed to all forms of censorship. Its Web site contains a large amount of information on all aspects of library operation, including news and opinions about legislation that affects Internet access at libraries.

GetNetWise
(202) 638-4370
e-mail: dyates@getnetwise.org
Web site: www.getnetwise.org

GetNetWise is a public service sponsored by a broad-based coalition of companies, public interest organizations, nonprofits and trade associations committed to empowering Internet users with the tools they need to keep their interact experience positive, safe and secure. Its Web site contains a safety guide and links to tools for families such as filtering and blocking software.

i-SAFE
5900 Pasteur Court, Suite 100, Carlsbad, CA   92008

(760) 603-7911 • fax: (760) 603-8382

Web site: www.i-safe.org

Endorsed by the U.S. Congress, i-SAFE is a nonprofit foundation whose mission is to educate and empower youth to make their Interact experiences safe and responsible. The goal is to educate students on how to avoid dangerous, inappropriate, or unlawful online behavior, which it accomplishes through K-12 curriculum and community outreach programs to parents, law enforcement, and community leaders. Its Web site contains the i-LEARN Online educational program, which requires registration and information about other offerings.

Internet Keep Safe Coalition

5220 36th Street North, Arlington, VA   22207

(703) 536-1637 • fax: (703) 536-3052

e-mail: info@ikeepsafe.org

Web site: www.ikeepsafe.org

The Internet Keep Safe Coalition is a nonprofit organization that teaches basic rules of Internet safety to children and parents. Its Web site contains material for children and for educators, as well as a complete children's book, *Faux Paw the Techno Cat: Adventures on the Internet.*

Mobilizing America's Youth (MAY)

1133 19th St. NW, Floor 9, Washington, DC   20036

(202) 736-5703 • (fax) 202-659.3716

Web site: www.mobilize.org

MAY is an all-partisan network dedicated to educating, empowering, and energizing young people to increase civic engagement and political participation. Among its projects is Save Our Social Networks, which aims to defeat the Deleting Online Predators Act (DOPA) or any similar legislation that would require social networking sites to be blocked by schools and libraries. Its Web site contains news and articles about pending legislation.

National Center for Missing & Exploited Children (NCMEC)

699 Prince Street, Alexandria, VA   22314-3175
(703) 274-3900 • fax: (703) 274-2200
Web site: www.missingkids.com

NCMEC is a private, nonprofit organization that helps prevent child abduction and sexual exploitation, helps find missing children, and assists victims of child abduction and sexual exploitation as well as their families and the professionals who serve them. Its Web site offers resources for parents, educators, and teens, such as NetSmartz411 (www.netsmartz411.org), a searchable knowledge base about Internet safety with an "Ask the Expert" option. It also operates a CyberTipline for reporting cases of online sexual exploitation of children.

Project Safe Childhood (PSC)

Web site: www.projectsafechildhood.gov

PSC is a program of the U.S. Department of Justice that aims to combat the proliferation of technology-facilitated sexual exploitation crimes against children. Its Web site offers a video, speeches by government officials, and government publications related to Internet safety.

Staysafe.org

Web site: www.staysafe.org

Staysafe.org is an educational site funded by Microsoft that is intended to help consumers understand the positive aspects of the Internet as well as how to manage a variety of safety and security issues that exist online. It contains articles such as "What the Social Web Looks Like to Parents" for teenagers, as well as information for parents, educators, and other adults.

Web Wise Kids

P.O. Box 27203, Santa Ana, CA   92799
(866) 932-9473 • fax: (714) 435-0523
e-mail: info@webwisekids.org

Web site: www.webwisekids.org

Web Wise Kids is a nonprofit online safety organization that works with kids, parents, teachers, and law enforcement to protect today's youth from online dangers. It offers educational software for classroom use, plus an interactive, multimedia computer game, *Missing*—designed to show, rather than tell, kids about predators who use the Internet—which can be purchased from its Web site.

Wired Safety
e-mail: webmaster@wiredsafety.org
Web site: www.wiredsafety.org

WiredSafety is a nonprofit online safety, education, and help group. It has more than 9,000 volunteers worldwide who help online victims of cybercrime and harassment, assist law enforcement to prevent and investigate cybercrimes, and provide information on all aspects of online safety, privacy, and security. It operates CyberLawEnforcement (CyberCops.org), StopCyberbullying (Cyberbullies.org), Teenangels.org, InternetSuperHeroes.org, and KatiesPlace.org, as well as WiredSafety.org.

Bibliography

Books

Dan Appleman   *Always Use Protection: A Teen's Guide to Safe Computing.* Berkeley, CA: Apress, 2004.

David Buckingham and Rebekah Willett, eds.   *Digital Generations: Children, Young People, and the New Media.* Mahwah NJ: Lawrence Erlbaum Associates, 2006.

Peter Buckley   *The Rough Guide to MySpace & Online Communities.* London and New York: Rough Guides, 2007.

Chap Clark and Dee Clark   *Disconnected: Parenting Teens in a MySpace World.* Grand Rapids, MI: Baker Books, 2007.

Linda Criddle and Nancy Muir   *Look Both Ways: Help Protect Your Family on the Internet.* Redmond, WA: Microsoft Press, 2006.

Laney Dale   *A Parent's Guide to MySpace.* Redondo Beach, CA: DayDream Publishers, 2006.

W. D. Edmiston   *Why Parents Should Fear MySpace.* Longwood, FL: Xulon Press, 2007.

Kevin M. Farnham and Dale G. Farnham   *MySpace Safety: 51 Tips for Teens and Parents.* Pomfret, CT: How-to Primers, 2006.

Dan Gillmor	*We the Media: Grassroots Journalism by the People, for the People.* Sebastopol, CA: O'Reilly, 2006.
Rebecca Hagelin	*Home Invasion: Protecting Your Family in a Culture that's Gone Stark Raving Mad.* Nashville, TN: Nelson, 2005.
Chris Hansen	*To Catch a Predator: Protecting Your Kids from Online Enemies Already in Your Home.* New York: Dutton, 2007.
David H. Holtzman	*Privacy Lost: How Technology Is Endangering Your Privacy.* San Francisco: Jossey-Bass, 2006.
Ryan Hupfer, Mitch Maxson, and Ryan Williams	*MySpace for Dummies.* New York: Wiley, 2007.
Jason Illian	*MySpace, MyKids: A Parent's Guide to Protecting Your Kids and Navigating MySpace.com.* Eugene, OR: Harvest House, 2006.
Simon Johnson	*Keep Your Kids Safe on the Internet.* New York: McGraw-Hill, 2004.
Candice M. Kelsey	*Generation MySpace: Helping Your Teen Survive Online Adolescence.* New York: Marlowe, 2007.
Larry Magid and Anne Collier	*MySpace Unraveled: What It Is and How to Use It Safely.* Berkeley, CA: Peachpit Press, 2006.

Sharon R. Mazzarella, ed.	*Girl Wide Web: Girls, the Internet, and the Negotiation of Identity.* New York: Peter Lang, 2005.
Connie Neal	*MySpace for Moms and Dads: A Guide to Understanding the Risks and the Rewards.* Grand Rapids, MI: Zondervan, 2007.
Howard Rheingold	*Smart Mobs: The Next Social Revolution.* New York: Basic Books, 2002.
Daniel J. Solove	*The Digital Person: Technology and Privacy in the Information Age.* New York: New York University Press, 2006.
Katherine Tarbox	*A Girl's Life Online.* New York: Plume, 2004.
Nancy E. Willard	*Cyberbullying and Cyberthreats: Responding to the Challenge of Online Social Aggression, Threats, and Distress.* Champaign, IL: Research Press, 2007.
Nancy E. Willard	*Cyber-safe Kids, Cyber-savvy Teens: Helping Young People Learn to Use the Internet Safely and Responsibly.* San Francisco: Jossey-Bass, 2007.

Periodicals

Stephen Abram	"What Can MySpace Teach Us in School Libraries?" *Multimedia & Internet @ Schools,* July–August, 2006.

Julian Aiken	"On My Mind: Hands Off MySpace," *American Libraries*, August 2006.
Robert W. Ashmore	"Blocking MySpace from Your Space," *School Administrator*, October, 2006.
Anna Bahney	"Don't Talk to Invisible Strangers," *New York Times*, March 9, 2006.
Tara Bahrampour and Lori Aratani	"Teens' Bold Blogs Alarm Area Schools," *Washington Post*, January 17, 2006.
Theodora A. Blanchfield	"In Your Face," *Campaigns & Elections*, May 2006.
Kelly Carter	"Internet Terror," *Teen People*, June–July 2006.
Nolan Clay	"Death Row Inmates Use MySpace," *Daily Oklahoman*, January 21, 2007.
Alan Cohen	"Do You Know Where Your Kids Are Clicking?" *PC Magazine*, January 22, 2007.
Andrea DeSimone	"Are You Addicted to MySpace?" *Teen People*, September 2006.
Camille Dodero	"They Have MySpace in Heaven, Right?" *Boston Phoenix*, March 28, 2006.
Gretchen Dukowitz	"Out on MySpace, Then Out the Door: Think That No One Will Discover You're Gay on MySpace?" *Advocate*, June 6, 2006.

Bruce Einhorn and Olga Kharif — "China: Falling Hard for Web 2.0, Youngsters Are Flocking to Home-grown Versions of MySpace and You-Tube," *Business Week*, January 15, 2007.

Liz Else — "I'll Have to Ask My Friends," *New Scientist*, September 16, 2006.

Alan Finder — "Guess Who's Looking at Your Web Page: College Admissions Officials and Employers Are Starting to Check Out Candidates on Web Sites Like Facebook and MySpace," *New York Times Upfront*, September 18, 2006.

Caitlin Flanagan — "Babes in the Woods," *Atlantic Monthly*, July–August 2007.

Geoffrey H. Fletcher — "Power Up, Don't Power Down: Barring Students from Using Cell Phones, MySpace, and Other Communication Technologies Once They Enter the Classroom Is the Wrong Approach," *T H E Journal*, January 22, 2007.

Alison George — "Things You Wouldn't Tell Your Mother," *New Scientist*, September 16, 2006.

Jennifer Granick — "Face It: Privacy Is Endangered," *Wired News*, December 7, 2005.

Toddi Gutner — "MySpace for the Sandlot Set: Social Networks Like Club Penguin Are Quickly Catching Pre-tweens' Attention," *Business Week*, October 2, 2006.

Jessi Hempel	"The MySpace Generation: They Live Online. They Buy Online. They Play Online. Their Power Is Growing," *Business Week*, December 12, 2005.
Bill Hewitt	"MySpace Nation: The Controversy," *People*, June 5, 2006.
Janet Kornblum	"Meet My 5,000 New Best Pals," *USA Today*, September 19, 2006.
Sarah Lacy	"Gather.com: Social Networking Grows Up," *Business Week Online*, October 26, 2006.
Laura Landro	"Social Networking Comes to Health Care," *Wall Street Journal*, December 27, 2006.
Paul Marks	"Keep Out of MySpace," *New Scientist*, June 10, 2006.
Irene E. McDermott	"I Need MySpace," *Searcher*, April 2006.
Scott Medintz	"Talkin' 'Bout MySpace Generation: Kids' Online Profiles Can Hurt Job Prospects Decades Down the Road," *Money*, February 1, 2006.
Wendy Melillo	"The Marines and MySpace," *Adweek*, November 20, 2006.
Tracy Mitrano	"A Wider World, Youth, Privacy, and Social Networking Technologies," *EDUCAUSE Review*, November–December 2006.

Connie Neal   "A Mom's Guide to MySpace: What
                     You Need to Know About This
                     Popular Website," *Today's Christian
                     Woman*, January–February 2007.

Mick O'Leary   "Ten Things You Don't Know About
                     MySpace," *Information Today*, Octo-
                     ber 2006.

Amanda Paulson   "Schools Grapple with Policing
                     Students' Online Journals," *Christian
                     Science Monitor*, February 2, 2006.

George H. Pike   "MySpace.com and Library Filters:
                     Deleting Online Predators Act." *Infor-
                     mation Today*, July–August 2006.

Kevin Poulson   "Scenes from the MySpace Backlash,"
                     Wired News, February 27, 2006.

Sean Rapacki   "Social Networking Sites: Why Teens
                     Need Places Like MySpace," *YALS*,
                     Winter 2007.

Julie Rawe   "How Safe Is MySpace?" *Time*, July 3,
                     2006.

Brock Read   "Think Before You Share: Students'
                     Online Socializing Can Have Unin-
                     tended Consequences," *Chronicle of
                     Higher Education*, January 20, 2006.

Tony Rehagen   "Virtual Hero," *Indianapolis Monthly*,
                     December 2006.

Karen Rile   "My MySpace Lesson," *Daughters*,
                     November—December 2006.

Andrew Romano   "Walking a New Beat: Surfing MySpace.com Helps Cops Crack the Case," *Newsweek*, April 24, 2006.

Patricia Sellers   "MySpace Cowboys," *Fortune*, September 4, 2006.

Noah Shachtman   "Murder on MySpace," *Wired*, December 2006.

William Stewart   "Private Lives Laid Bare on MySpace," *Times Educational Supplement*, August 18, 2006.

Brad Stone and Robbie Brown   "Web of Risks," *Newsweek*, August 28, 2006.

Julie Sturgeon   "Bullies in Cyberspace," *District Administration*, September 2006.

"Thousands Grieve for Anna, a Girl They Never Knew"   *Sydney Morning Herald*, June 11, 2006.

Andrew Trotter   "Social-Networking Web Sites Pose Growing Challenge for Educators," *Education Week*, February 15, 2006.

James Verini   "Will Success Spoil MySpace?" *Vanity Fair*, March 2006.

Thomas E. Wheeler   "Lessons from *The Lord of the Flies*: Protecting Students from Internet Threats and Cyber Hate Speech," *Journal of Internet Law*, July 2006.

Index